Variable Gain Control and Its Applications in Energy Conversion

The variable gain control method is a new construction technique for the control of nonlinear systems. By properly conducting state transformation that depends on the variable gains, the control design problem of nonlinear systems can be transformed into a gain construction problem, thus effectively avoiding the tedious iterative design procedure. Different from the classical backstepping method and forwarding design method, the structure of variable gain control is simpler in the sense that fewer design parameters are required, facilitating the improvement of system control performance.

To highlight the learning, research, and promotion of variable gain control, *Variable Gain Control and Its Applications in Energy Conversion* is written based on the research results of peers at home and abroad and combining the authors' latest research. This book presents innovative technologies for designing variable gain controllers for nonlinear systems. It systematically describes the origin and principles of variable gain control for nonlinear systems, focuses on the controller design and stability analysis, and reflects the latest research. In addition, variable gain control methods applied to energy conversion are also included.

Discussion remarks are provided in each chapter highlighting new approaches and contributions to emphasize the novelty of the presented design and analysis methods. In addition, simulation results are given in each chapter to show the effectiveness of these methods.

It can be used as a reference book or a textbook for students with some background in feedback control systems. Researchers, graduate students, and engineers in the fields of control, information, renewable energy generation, electrical engineering, mechanical engineering, applied mathematics, and others will benefit from this book.

Automation and Control Engineering
Series Editors - Frank L. Lewis, Shuzhi Sam Ge, and Stjepan Bogdan

System Modeling and Control with Resource-Oriented Petri Nets
MengChu Zhou, Naiqi Wu

Deterministic Learning Theory for Identification, Recognition, and Control
Cong Wang, David J. Hill

Optimal and Robust Scheduling for Networked Control Systems
Stefano Longo, Tingli Su, Guido Herrmann, Phil Barber

Electric and Plug-in Hybrid Vehicle Networks
Optimization and Control
Emanuele Crisostomi, Robert Shorten, Sonja Stüdli, Fabian Wirth

Adaptive and Fault-Tolerant Control of Underactuated Nonlinear Systems
Jiangshuai Huang, Yong-Duan Song

Discrete-Time Recurrent Neural Control
Analysis and Application
Edgar N. Sánchez

Control of Nonlinear Systems via PI, PD and PID
Stability and Performance
Yong-Duan Song

Multi-Agent Systems
Platoon Control and Non-Fragile Quantized Consensus
Xiang-Gui Guo, Jian-Liang Wang, Fang Liao, Rodney Swee Huat Teo

Classical Feedback Control with Nonlinear Multi-Loop Systems
With MATLAB® and Simulink®, Third Edition
Boris J. Lurie, Paul Enright

Motion Control of Functionally Related Systems
Tarik Uzunović, Asif Sabanović

Intelligent Fault Diagnosis and Accommodation Control
Sunan Huang, Kok Kiong Tan, Poi Voon Er, Tong Heng Lee

Nonlinear Pinning Control of Complex Dynamical Networks
Edgar N. Sanchez, Carlos J. Vega, Oscar J. Suarez, Guanrong Chen

Adaptive Control of Dynamic Systems with Uncertainty and Quantization
Jing Zhou, Lantao Xing, Changyun Wen

Robust Formation Control for Multiple Unmanned Aerial Vehicles
Hao Liu, Deyuan Liu, Yan Wan, Frank L. Lewis, Kimon P. Valavanis

Variable Gain Control and Its Applications in Energy Conversion
Chenghui Zhang, Le Chang, Cheng Fu

For more information about this series, please visit: https://www.crcpress.com/Automation-and-Control-Engineering/book-series/CRCAUTCONENG

Variable Gain Control and Its Applications in Energy Conversion

Chenghui Zhang
Le Chang
Cheng Fu

CRC Press
Taylor & Francis Group
Boca Raton London New York

CRC Press is an imprint of the
Taylor & Francis Group, an **informa** business

First edition published 2023
by CRC Press
6000 Broken Sound Parkway NW, Suite 300, Boca Raton, FL 33487-2742

and by CRC Press
4 Park Square, Milton Park, Abingdon, Oxon, OX14 4RN

CRC Press is an imprint of Taylor & Francis Group, LLC

ISBN: 978-1-032-49184-4 (hbk)
ISBN: 978-1-032-49272-8 (pbk)
ISBN: 978-1-003-39292-7 (ebk)

DOI: 10.1201/9781003392927

Typeset in Nimbus Roman font
by KnowledgeWorks Global Ltd.

Contents

Part II   Variable Gain Control for Feedforward Nonlinear Systems

Part III   *Variable Gain Control for Large-Scale Nonlinear Systems*

Part IV   Application of Variable Gain Control in Energy Conversion

Preface

In practical applications, a large number of physical systems are modeled by nonlinear dynamic equations, such as three-phase AC/DC power converters, permanent magnet synchronous motors, and smart microgrids. Compared to linear systems, nonlinear systems no longer satisfy the superposition principle and exhibit more complex properties, including the existence of multiple isolated equilibrium points and the finite-time escape of solutions. Due to the complex nature of nonlinear systems, the design of controllers is always a challenging task. Currently, most research focuses on nonlinear systems with specific structures. For example, high-gain control is often used for strict-feedback nonlinear systems, while low-gain control is suitable for feedforward nonlinear systems.

The variable gain control method is a new construction technique for controlling nonlinear systems. By properly performing a state transformation that depends on the variable gains, the control design problem of nonlinear systems can be transformed into a gain construction problem, effectively avoiding the tedious iterative design procedure. Unlike the classical backstepping method and the forwarding design method, the structure of variable gain control is simpler in the sense that fewer design parameters are required, which facilitates the improvement of system control performance.

To advance the learning, research and promotion of variable gain control, this book has been written based on the research findings of peers at home and abroad, combining our latest research. The results are presented in four parts. The first part deals with variable gain control for strict-feedback nonlinear systems (Chapters 2–4). The second part deals with the design of variable gain control for feedforward nonlinear systems (Chapters 5–7). After that, the third part deals with variable gain control for large-scale nonlinear systems (Chapters 8–10). Parts 1–3 mainly focus on theoretical research and have remarkable features in increasing gain feedback control (Chapter 2), output feedback control (Chapter 3), fixed time stabilizing control (Chapter 4), decreasing gain feedback control (Chapter 5), asymptotic stabilization of time-delayed systems (Chapter 6), control design for discrete-time systems (Chapter 7), decentralized control of interconnected nonlinear systems (Chapter 8), distributed control of multi-agent systems (Chapter 9) and consensus control of time-delayed systems (Chapter 10). The advantages of the variable gain method in improving system performance are demonstrated through theoretical analysis and simulation examples.

In the fourth part, the developed variable gain control schemes are used to solve control problems in the field of energy conversion. In Chapter 11, static gain control and variable gain control strategies for three-phase AC/DC power converters are proposed; in Chapter 12, the fixed-time speed tracking control of a permanent magnet synchronous motor via variable gain control is discussed; in Chapter 13, a distributed robust secondary control for island microgrid is developed, which aims to compensate for voltage and frequency deviations. In summary, the above application examples further demonstrate the applicability and advantages of variable gain control methods.

Each chapter includes discussion notes highlighting new approaches and contributions to emphasize the novelty of the design and analysis methods presented. Simulation results are also given in each chapter to show the effectiveness of these methods.

The book can be used as a reference or textbook for students with prior knowledge of feedback control. Researchers, graduate students, and engineers in control, information, renewable energy generation, electrical engineering, mechanical engineering, applied mathematics, and others will benefit from this book. We are deeply indebted to our beloved families who enabled us to write this book. We are grateful for the helpful suggestions from our colleagues Prof. Xianfu Zhang, Prof. Lantao Xing, Prof. Kailong Liu, Prof. Yan Li, Dr. Hanfeng Li, and Dr. Xiandong Chen. Finally, the authors are grateful to Shandong University, China, for providing numerous resources for our research work. The research presented in this book was supported in part by the Innovation Research Group of National Natural Science Foundation of China (NSFC) under Grant 61821004 and in part by the Key Program of Automobile Joint Fund of NSFC under Grant U1964207.

Part I

Variable Gain Control for Strict-Feedback Nonlinear Systems

1 Introduction

1.1 BACKGROUND

In recent decades, there has been tremendous research interest in the control of nonlinear systems, fueled by the increasingly stringent control requirements of modern systems/devices. For example, to effectively combat the energy crisis, countries around the world are vigorously developing electric vehicles. To improve the durability and safety of electric vehicles, battery pack control and management is one of the most important problems to be solved [114]. The charging and discharging of batteries is a complex process involving the conversion of electrical, chemical and thermal energy [88, 145] and essentially involves nonlinear and uncertain dynamics. Therefore, nonlinear systems can better capture the nature of practical control systems compared to linear systems.

However, nonlinear systems often suffer from uncertain parameters/dynamics and modeling errors. This is due to a number of factors, such as the inherent interaction between the control system and the environment, the lack of understanding of the operating law of the control plant, and the limited precision of the measurement tools available. These uncertainties and modeling errors can seriously affect the system control performance. In addition, nonlinear systems have many unique properties, such as the existence of multiple isolated equilibrium points, the phenomenon of finite-time escape, or the non-uniqueness of solutions. Therefore, the control of nonlinear systems is never a trivial task.

Among all available nonlinear control methods, variable gain control stands out for its ability to deal with system uncertainties. It can be divided into three categories: static high-gain control, static low-gain control and dynamic gain control. The earliest result of variable gain control comes from the past decades, where asymptotic stabilization was solved for strict-feedback nonlinear systems with known increment rate [102]. Since then, many remarkable results have been presented. In the meantime, this type of control technique has also been applied to many practical systems, such as the one-link manipulator systems and the chemical reactor systems. Nevertheless, there are still many open problems that remain unsolved. This book aims to give readers a comprehensive introduction to the latest theoretical and application results in this field.

1.2 RESEARCH STATUS

The variable gain control method is mainly used to stabilize nonlinear systems with strict-feedback and feedforward structures. Moreover, it can also solve the control design problem for large-scale nonlinear systems. This method has the advantages of high efficiency and simplicity, thus it has great potential for control design in many practical applications.

1.2.1 VARIABLE GAIN CONTROL FOR STRICT-FEEDBACK NONLINEAR SYSTEMS

Strict-feedback nonlinear systems represent a wide range of nonlinear systems, and an effective control method for such systems is backstepping [59]. The idea of backstepping is to select a suitable Lyapunov function in different design steps and then combine them into a final Lyapunov function. Based on backstepping, the problem of global asymptotic stabilization by output feedback for strict-feedback nonlinear systems is considered in [102]. In [44], a strict-feedback system subject to time delay is examined. Although backstepping is widely used in controlling strict-feedback nonlinear systems, the huge increase in computational complexity is unacceptable when the system size is large.

Variable high-gain control is another construction technique for studying nonlinear systems with a strict-feedback structure. By properly choosing coordinate transformations that depend on static or variable gains, the control design problem for nonlinear systems can be transformed into an indefinite gain construction problem [102], thus effectively avoiding the tedious iterative design procedure. In recent years, variable gain control has been applied to strict-feedback systems where nonlinear terms satisfy different growth conditions such as the state-dependent growth condition [48, 103]. Despite the above progress, the corresponding work for more complex systems is still limited, especially when time-varying dynamics and time delays are involved.

Fixed-time stabilizing control (FSC) has been extensively studied over the last decade. This method ensures that the system trajectories converge to zero within a fixed time for any initial conditions [101]. In [117] and [99], a time-varying function is used to regulate the rate of convergence to achieve stabilization with fixed time, but the control law becomes invalid after the specified settling time. Recently, [121] studied the fixed-time fuzzy tracking control problem for a class of unknown nonlinear systems. However, the proposed method suffers from the problem of computational complexity caused by the backstepping design. Due to its merits, variable gain control can provide an effective solution for stabilizing fixed-time stabilization without the above problems.

1.2.2 VARIABLE GAIN CONTROL FOR FEEDFORWARD NONLINEAR SYSTEMS

On the other hand, the control design of feedforward nonlinear systems has been extensively studied because they can model many practical systems, such as planar vertical take-off and landing aircraft and the inertia wheel pendulum. Generally, two types of conditions are used for feedforward nonlinear systems to perform stability analysis. The first is that the nonlinear system dynamics is bounded by the sum of the quadratic or higher terms of its arguments. Under this condition, the stabilization problem is studied in [122, 95, 136, 161] by the saturation control method. The second one is that the nonlinear dynamics satisfy the Lipschitz condition. This condition can involve a coefficient in various forms, such as a known constant [152], an unknown constant [9], a function dependent on the input or output [147, 58], and a function dependent on both the input and output [54].

Low-gain feedback control is an effective method for feedforward nonlinear systems. Although several low-gain feedback control methods [74, 75] have been developed to achieve various control goals, the development of low-gain feedback control has attracted less attention for feedforward nonlinear systems and only a few results are obtained [137]. For feedforward nonlinear systems, the time-varying terms are always involved in the nonlinear functions whose growth rate increases with time, but the existing results dealing with bounded nonlinear functions cannot be applied. In practical systems, there are often time delays because of the transmission of the measurement signals. The problem of asymptotic stabilization is solved in [94] for feedforward nonlinear systems with input delay. However, the recursive design method cannot be directly extended to feedforward nonlinear systems with time delays, as the construction of a Krasovskii function or a Razimikhin function is non-trivial. This means that new control design methods are required for the control of feedforward nonlinear systems with time delays. Furthermore, it should be mentioned that the corresponding results for discrete-time feedforward nonlinear systems are also limited [135, 1, 92].

1.2.3 VARIABLE GAIN CONTROL FOR LARGE-SCALE NONLINEAR SYSTEMS

The last decade has seen a resurgence of interest in the analysis and synthesis of large-scale dynamical systems consisting of a large number of spatially distributed subsystems (or agents, units). These systems include power systems, water/gas distribution networks, transportation networks, multi-robot systems, wireless sensor networks and unmanned factories [35, 17, 32]. An obvious feature of such systems is that the subsystems are closely *interconnected* (or coupled) with each other. Up to now, some efforts have been made in the control design of a wide variety of interconnected systems, see, e.g., [56, 58, 110, 134, 31, 70, 156, 20]. Recently, the dynamic high-gain control approach is generalized in [151, 155, 82, 153] to stabilize several classes of strict-feedback and feedforward large-scale nonlinear systems. However, such a centralized dynamic gain scheduling method proposed in [151, 155, 82, 153] essentially suffers from computational complexity and communication burden issues, especially when the system scale grows larger.

The multi-agent system consists of several interactive agents. A crucial problem in the distributed control of multi-agent systems is the consensus problem, i.e., the development of suitable protocols and algorithms so that the group of agents can reach consensus via a common communication network. Many works have studied this problem, such as the consensus of linear multi-agent systems [113, 125, 90], second-order agent systems [41, 163, 68, 107], multi-agent nonlinear systems with Lipschitz non-linearity [71], and strict-feedback nonlinear multi-agent systems [127, 141, 144]. However, the variable gain control results for feedforward nonlinear multi-agent and strict-feedback nonlinear multi-agent systems with time delays are still limited.

1.2.4 VARIABLE GAIN CONTROL IN ENERGY CONVERSION

Countries around the world have already set 2030 and 2050 targets to reduce carbon emissions and curb global warming. One of the most important technologies for achieving these goals is renewable energy generation [6]. In the generation and use of renewable energy, energy conversion devices are essential to enable people to use energy conveniently and effectively. These devices include power converters, permanent magnet synchronous motors, etc. It should be mentioned that these devices are essentially nonlinear systems whose control performance is easily affected by numerous uncertainties. It is also obvious that the increasing penetration of renewable energies is leading to problems with system stability. Therefore, new control technologies are increasingly needed to ensure system stability, reliability and safety in energy conversion. [89].

In recent years, many advanced control techniques have been developed for energy conversion, such as passive control [108, 24, 39], sliding mode control [79, 104, 2], and backstepping control [124, 120]. Among these methods, sliding mode control offers robustness to system uncertainties, but its inherent chattering problem is not acceptable for power systems. Backstepping control divides the high-order system into several subsystems by introducing several virtual control laws so that the actual controller can be designed in a recursive manner [142]. However, the stability analysis of backstepping control is based on an accurate mathematical model and is therefore invalid if system uncertainties exist. Since variable gain control has proven to be a promising tool for controlling uncertain nonlinear systems, its applications in energy conversion offer great potential for solving energy conversion problems.

1.3 STATIC GAIN CONTROL DESIGN OF NONLINEAR SYSTEMS

To give readers insight into variable gain control, we recall two basic methods of gain control below, namely static high-gain control and static low-gain control. More details can be referred to [103, 56, 55, 77, 152, 137, 51, 76, 122, 105, 91, 49].

Consider the following nonlinear system:

$$\begin{cases} \dot{x}_j = x_{j+1} + f_j(\cdot), & j = 1, 2, \ldots, n-1, \\ \dot{x}_n = u + f_n(\cdot), \end{cases} \tag{1.1}$$

where the nonlinear functions $f_i(\cdot)$, $i = 1, 2, \ldots, n$, are continuous.

If $f_j(\cdot)$ is in the form of $f_j(t, x_{j+2}, x_{j+3}, \ldots, x_n, u)$, $j = 1, 2, \ldots, n-2$, $f_{n-1}(u)$ and $f_n(\cdot) = 0$, the system (1.1) is called a feedforward nonlinear system; If $f_j(\cdot)$ has the form $f_j(t, x_1, x_2, \ldots, x_j)$, $j = 1, 2, \ldots, n$, then (1.1) is called a strict-feedback nonlinear system.

The backstepping and forwarding method can not only construct a robust feedback controller, but also provide a Lyapunov function to adapt system states to specific control requirements. However, the backstepping and forwarding method suffers from a tedious controller construction process. The variable gain design method is a new design technique for the study of strict-feedback and feedforward nonlinear systems. By properly choosing the state transformation that depends on static or variable gains, the control problem of nonlinear systems can be transformed into an indefinite gain construction problem, effectively avoiding the tedious iterative design procedure. When using the variable gain design method, the nonlinear functions usually satisfy the following assumptions:

Assumption 1.1 (Feedforward nonlinear systems). *For any* $x = (x_1, x_2, \ldots, x_n)^T \in \mathbb{R}^n$, $u \in \mathbb{R}$, *there exists a function or constant* $g(\cdot)$ *such that*

$$|f_i(x_{i+2}, x_{i+3}, \ldots, x_n, u)| \leq g(\cdot)(|x_{i+2}| + |x_{i+3}| + \ldots + |x_n| + |u|)$$

holds for any $i = 1, 2, \ldots, n-2$ *and* $|f_{n-1}(u)| \leq g(\cdot)|u|$.

Assumption 1.2 (Strict-feedback nonlinear systems). *For any* $x = (x_1, x_2, \ldots, x_n)^T \in \mathbb{R}^n$, *there exists a function or constant* $g(\cdot)$ *such that*

$$|f_i(x_1, x_2, \ldots, x_i)| \leq g(\cdot)(|x_1| + |x_2| + \ldots + |x_i|)$$

holds for any $i = 1, 2, \ldots, n$.

The static high-gain feedback control and static low-gain feedback control are given below based on $g(\cdot) = c$, where c is a known positive constant.

1.3.1   STATIC HIGH-GAIN FEEDBACK CONTROL

The static high-gain feedback control is used to design a control law with a parameter to be determined. This parameter can change the control gains, and adjust the convergence rate. Moreover, this parameter can effectively handle the influence of nonlinear functions.

Here we consider the nonlinear system with strict-feedback (1.1) under Assumption 1.2, where $g(\cdot) = c$ is a known positive constant. The high-gain feedback control is designed as

$$u = -h^n k_1 x_1 - h^{n-1} k_2 x_2 - \ldots - h k_n x_n, \tag{1.2}$$

where $k_1, k_2, \ldots, k_n$, are the constants such that the polynomial

$$s^n + k_n s^{n-1} + \ldots + k_2 s + k_1$$

is Hurwitz; that is, its roots have negative real parts. It should be pointed out that h is a positive constant to regulate the control gains.

Consider the state transformation

$$z_1 = \frac{x_1}{h}, z_2 = \frac{x_2}{h^2}, \ldots, z_n = \frac{x_n}{h^n},$$

and we can get

$$
\begin{cases}
\dot{z}_1 = hz_2 + \dfrac{1}{h} f_1(x_1), \\[2mm]
\dot{z}_2 = hz_3 + \dfrac{1}{h^2} f_2(x_1, x_2), \\[2mm]
\dot{z}_3 = hz_4 + \dfrac{1}{h^3} f_3(x_1, x_2, x_3), \\[2mm]
\quad\vdots \\[2mm]
\dot{z}_n = \dfrac{u}{h^n} + \dfrac{1}{h^n} f_n(x_1, x_2, x_3, \ldots, x_n),
\end{cases}
$$

where $z = (z_1, z_2, \ldots, z_n)^T \in \mathbb{R}^n$ is the state.

The control (1.2) is converted into

$$
u = -h^{n+1} \left(k_1 z_1 + k_2 z_2 + \ldots + k_n z_n \right).
$$

Let

$$
A = \begin{pmatrix}
0 & 1 & 0 & \cdots & 0 \\
0 & 0 & 1 & \cdots & 0 \\
\vdots & \vdots & \vdots & & \vdots \\
0 & 0 & 0 & \cdots & 1 \\
0 & 0 & 0 & \cdots & 0
\end{pmatrix}, B = \begin{pmatrix}
0 \\
0 \\
\vdots \\
0 \\
1
\end{pmatrix}, F = \begin{pmatrix}
\frac{1}{h} f_1(x_1) \\
\frac{1}{h^2} f_2(x_1, x_2) \\
\cdots \\
\frac{1}{h^{n-1}} f_{n-1}(x_1, \ldots, x_{n-1}) \\
\frac{1}{h^n} f_n(x_1, x_2, \ldots, x_n)
\end{pmatrix},
$$

and

$$
K = \begin{pmatrix} k_1 & k_2 & \cdots & k_n \end{pmatrix}.
$$

Then, the closed-loop system can be rewritten as

$$
\dot{z} = h(A - BK)z + F. \tag{1.3}
$$

Since the elements of K are the coefficients of the Hurwitz polynomial, we know that $A - BK$ is the Hurwitz matrix. Then, there is a positive defined matrix P such that

$$
(A - BK)^T P + P(A - BK) \leq -I.
$$

Consider the Lyapunov function candidate

$$
V = z^T P z,
$$

and its time derivative is computed as

$$
\begin{aligned}
\dot{V}|_{(1.3)} &= \left(hz^T (A - BK)^T + F^T \right) Pz + z^T P \left(h(A - BK)z + F \right) \\
&= hz^T \left((A - BK)^T P + P(A - BK) \right) z + 2z^T PF \\
&\leq -h\|z\|^2 + 2z^T PF.
\end{aligned} \tag{1.4}
$$

From Assumption 1.2 and $h \geq 1$, we get

$$
\begin{aligned}
\left|\frac{1}{h^i} f_i(x_1, x_2, \ldots, x_i)\right| &\leq \frac{1}{h^i} c \left(|x_1| + |x_2| + \ldots + |x_i|\right) \\
&\leq \frac{1}{h^i} c \left(|hz_1| + |h^2 z_2| + \ldots + |h^i z_i|\right) \\
&\leq c \left(\frac{1}{h^{i-1}} |z_1| + \frac{1}{h^{i-2}} |z_2| + \ldots + |z_i|\right) \\
&\leq c \left(|z_1| + |z_2| + \ldots + |z_i|\right) \\
&\leq c\sqrt{n} \|z\|.
\end{aligned}
$$

The estimation of F is
$$
\|F\| \leq cn \|z\|.
$$

Back to (1.4), we have

$$
\dot{V}|_{(1.3)} \leq -h\|z\|^2 + 2\|P\| \|z\| \|F\| \leq -h\|z\|^2 + 2nc\|P\| \|z\|^2.
$$

Let
$$
h \geq \max\{4nc\|P\|, 1\},
$$

and we get
$$
\dot{V}|_{(1.3)} \leq -\frac{h}{2} \|z\|^2.
$$

Thus, we get the convergence of z. Since h is a constant, we can further deduce the convergence of x.

1.3.2   STATIC LOW-GAIN FEEDBACK CONTROL

Similar to high-gain static feedback control for strict feedback nonlinear systems, low-gain static feedback control is given for feedforward nonlinear systems. Low-gain static feedback control introduces a control law with a parameter to be determined that can change the control gains and adjust the convergence rate.

Nonlinear feedforward systems that have an upper triangular form, are widely studied by designing a low-gain static feedback control. Here we consider the feedforward nonlinear system (1.1) under Assumption 1.1, where $g(\cdot) = c$ is a known positive constant.

The low-gain feedback control is designed as

$$
u = -\frac{k_1}{h^n} x_1 - \frac{k_2}{h^{n-1}} x_2 - \ldots - \frac{k_n}{h} x_n, \tag{1.5}
$$

where $k_1, k_2, \ldots, k_n$, are the constants such that the polynomial

$$
s^n + k_n s^{n-1} + \ldots + k_2 s + k_1
$$

is Hurwitz.

Consider the state transformation

$$z_1 = \frac{x_1}{h^n}, z_2 = \frac{x_2}{h^{n-1}}, \ldots, z_n = \frac{x_n}{h},$$

and we can get

$$
\begin{cases}
\dot{z}_1 = \frac{1}{h}z_2 + \frac{1}{h^n}f_1(x_3, x_4, \ldots, x_n, u), \\
\dot{z}_2 = \frac{1}{h}z_3 + \frac{1}{h^{n-1}}f_2(x_4, \ldots, x_n, u), \\
\vdots \\
\dot{z}_{n-1} = \frac{1}{h}z_n + \frac{1}{h^2}f_{n-1}(u), \\
\dot{z}_n = \frac{u}{h},
\end{cases}
$$

where $z = (z_1, z_2, \ldots, z_n)^T \in \mathbb{R}^n$ is the state.

The control (1.5) is converted into

$$u = -(k_1 z_1 + k_2 z_2 + \ldots + k_n z_n).$$

Let

$$
A = \begin{pmatrix}
0 & 1 & 0 & \cdots & 0 \\
0 & 0 & 1 & \cdots & 0 \\
\vdots & \vdots & \vdots & & \vdots \\
0 & 0 & 0 & \cdots & 1 \\
0 & 0 & 0 & \cdots & 0
\end{pmatrix}, B = \begin{pmatrix} 0 \\ 0 \\ \vdots \\ 0 \\ 1 \end{pmatrix}, F = \begin{pmatrix} \frac{1}{h^n}f_1(x_3, x_4, \ldots, x_n, u) \\ \frac{1}{h^{n-1}}f_2(x_4, \ldots, x_n, u) \\ \cdots \\ \frac{1}{h^2}f_{n-1}(u) \\ 0 \end{pmatrix},
$$

and

$$K = \begin{pmatrix} k_1 & k_2 & \cdots & k_n \end{pmatrix}.$$

Then, the closed-loop system can be rewritten as

$$\dot{z} = \frac{1}{h}(A - BK)z + F. \tag{1.6}$$

Since the elements of K are coefficients of a Hurwitz polynomial, we get $A - BK$ is the Hurwitz matrix. Then, there is a positive definite matrix P such that

$$(A - BK)^T P + P(A - BK) \le -I.$$

Consider the Lyapunov function candidate

$$V = z^T P z,$$

and its time derivative is computed as

$$
\begin{aligned}
\dot{V}|_{(1.6)} &= \left(\frac{1}{h}z^T(A - BK)^T + F^T \right) Pz + z^T P \left(\frac{1}{h}(A - BK)z + F \right) \\
&= \frac{1}{h}z^T \left((A - BK)^T P + P(A - BK) \right) z + 2z^T PF \\
&\le -\frac{1}{h}\|z\|^2 + 2z^T PF.
\end{aligned}
$$

From Assumption 1.1 and $h \geq 1$, we get for $i = 1, 2, \ldots, n-2$,

$$\left| \frac{1}{h^{n+1-i}} f_i(x_{i+2}, x_{i+3}, \ldots, x_n, u) \right|$$

$$\leq \frac{c}{h^{n+1-i}} \left(|x_{i+2}| + |x_{i+3}| + \ldots + |x_n| + |u| \right)$$

$$\leq \frac{c}{h^{n+1-i}} \left(|h^{n-1-i}z_{i+2}| + |h^{n-2-i}z_{i+3}| + \ldots + |hz_n| + |Kz| \right)$$

$$\leq \frac{c}{h^2} \left(|z_{i+2}| + \frac{1}{h}|z_{i+3}| + \ldots + \frac{1}{h^{n-i-2}}|z_n| + |Kz| \right)$$

$$\leq \frac{c}{h^2} \left(|z_{i+2}| + |z_{i+3}| + \ldots + |z_n| + \|K\|\|z\| \right)$$

$$\leq \frac{c}{h^2} \left(\sqrt{n} + \|K\| \right) \|z\|,$$

and $|\frac{1}{h^2} f_{n-1}(u)| \leq \frac{c}{h^2}\|K\|\|z\|$.
The estimation of F is

$$\|F\| \leq \frac{c}{h^2}(n + \sqrt{n}\|K\|)\|z\|.$$

Back to (1.4), we have

$$\dot{V}|_{(1.6)} \leq -\frac{1}{h}\|z\|^2 + 2\|P\|\|z\|\|F\| \leq -\frac{1}{h}\|z\|^2 + \frac{2}{h^2}c(n + \sqrt{n}\|K\|)\|P\|\|z\|^2.$$

Let

$$h \geq \max\{4c(n + \sqrt{n}\|K\|)\|P\|, 1\},$$

and we get

$$\dot{V}|_{(1.6)} \leq -\frac{1}{2h}\|z\|^2.$$

Thus, we get the convergence of z. Since h is a constant, we can further deduce the convergence of x.

It should be noted that the static high-/low-gain design method can only handle cases where $g(\cdot)$ is a known positive constant. However, when $g(\cdot)$ is an unknown positive constant, a time-varying function, or a known function that depends on system signals, the static high-/low-gain design method becomes invalid, which promotes the development and proposal of the time-varying/dynamic gain design method.

1.4 OBJECTIVES

The main objectives of this book are as follows:

- The first part of this book presents the variable gain control method for strict-feedback nonlinear systems. A control method with an increasing gain will be proposed for a class of strict-feedback nonlinear systems. We will also develop new output feedback control and fixed-time feedback control schemes for two types of strict-feedback nonlinear systems.

- In the second part of this book, the variable gain control method is proposed for stabilization control of feedforward nonlinear systems. A feedback control strategy with a decreasing gain will be established, and the asymptotic stabilization control and discrete-time control schemes for different classes of feedforward nonlinear systems are presented.
- The third part of this book uses the variable gain control method for large-scale nonlinear systems. New decentralized control strategies for different types of feedforward large-scale systems with global dynamic gains and distributed gains are presented. New distributed control methods for feedforward multi-agent systems are proposed. We also propose a new distributed output feedback control strategy for consensus control of strict-feedback nonlinear multi-agent systems subject to time delays.
- The fourth part of this book applies the variable gain control method to three types of energy conversion systems. These three types of systems have different nonlinear properties, so we will consider each of them separately in the designs of the control strategies. The dynamic gain control method will be applied to the control of three-phase AC/DC power converters. We will propose a fixed-time control strategy for the speed regulation of permanent magnet synchronous motors using the variable gains control method. We will also present a distributed robust secondary control scheme for islanded microgrids.

1.5 PREVIEW OF CHAPTERS

This book aims to address the above challenges and develop new variable gain design methods. On the one hand, we present the basic design methods for both strict-feedback nonlinear systems and feedforward nonlinear systems. On the other hand, we apply these design methods to the study of electronic systems. The preview of each chapter is given as follows:

Chapter 1 gives a brief overview of the history and different types of feedback control of nonlinear systems. It then explains the basic ideas behind the development of high-/low-gain control methods and presents the objectives in the study of these methods.

Chapter 2 develops the control for the system with complex nonlinearities. Our main goal is to answer why we develop the variable gain design method and which nonlinear conditions can be handled by the variable gain design method. Thus, a control method with an increasing gain is designed to stabilize strict-feedback nonlinear systems with complex and time-varying nonlinear dynamics.

Chapter 3 uses the increasing gain feedback control method to achieve the output feedback control for a class of strict-feedback nonlinear systems. Based on a dynamic high-gain observer, a delay-independent controller is constructed to make the closed-loop system asymptotically stable.

Chapter 4 uses the variable gain design method to improve the performance of the system. We design a fixed-time feedback control for nonlinear systems, and the state of the generated closed-loop system can reach zero within a fixed time period.

Chapter 5 introduces feedback control with a decreasing gain, which aims to stabilize feedforward nonlinear systems. A controller with a dynamic gain is designed and the stability of the original system is rigorously analyzed.

Chapter 6 applies variable gain control to the asymptotic stabilization of a class of feedforward time-delay systems. A state feedback controller is proposed which has a very simple structure and is therefore easy to implement in practice.

Chapter 7 introduces discrete-time feedforward nonlinear systems and proposes an asymptotic stabilization control strategy. A modified control method with a dynamic gain is proposed to achieve global stabilization.

Chapter 8 introduces the design method for large-scale nonlinear systems. We first consider decentralized control to achieve local stabilization. Then, we design a stabilizing controller with a global dynamic parameter. Finally, distributed parameter regulation is introduced and we propose a method for global stabilization.

Chapter 9 introduces the variable gain design method for multi-agent systems. We first consider the consensus problem by designing the feedback control with a low gain. Then a time-varying gain is designed to handle the case where the nonlinear terms are complex. It is shown that the variable gain control method can be used to design the consensus protocol for multi-agent systems.

Chapter 10 applies the variable gain design method to multi-agent systems with time delays. We first consider the consensus problem by designing the feedback control with a high gain. Secondly, the distributed output feedback control strategy is proposed by designing a time-varying gain. It is shown that the variable gain control method can be used to design the consensus protocol for multi-agent systems subject to time delays.

Chapter 11 applies static gain control and dynamic gain control to the three-phase AC/DC converter with the aim of creating simple and robust control schemes. The dynamic mathematical model of the AC/DC converter is built in the $\alpha\beta$ framework, with the introduced new control signals supporting the independent regulation of active and reactive power without a phase-locked loop. The comparative study discusses the advantages of dynamic gain control for the AC/DC converter.

In Chapter 12, a fixed-time dynamic gain control for the permanent magnet synchronous motor (PMSM) is proposed to ensure dynamic performance and control accuracy. A dynamic state transformation is introduced to transform the original PMSM system into a system whose control coefficients are regulated by two dynamic parameters. The fixed-time controller and the dynamic parameters are designed using the transformed state variables. Case studies verify the effectiveness and advantages of the proposed fixed-time method for PMSM under different working conditions.

Chapter 13 proposes a disturbance observer-based distributed robust secondary control for the islanded microgrid via the static gain design procedure. The system uncertainties are grouped and taken as a lumped disturbance, and a super-twisting disturbance observer is proposed to estimate this lumped disturbance. Theoretical analysis and case studies demonstrate the effectiveness of the proposed distributed secondary control method.

Finally, Chapter 14 concludes the whole book by summarizing the main approaches and contributions and discussing some promising open problems.

2 Global State Feedback Control for Strict-Feedback Nonlinear Systems

This chapter considers the global state feedback control problem for strict-feedback nonlinear systems. It is assumed that the uncertain nonlinearities are bounded by a time-varying function and a continuous function of the system states. Based on the Lyapunov stability theorem and the variable gain method, a state feedback controller with two gains is designed. The dynamic gain, which depends on the system states, is designed to deal with the nonlinearities of the complex system, and the time-varying gain is constructed to deal with the influences of the resulting time-varying functions. Unlike many existing control methods for nonlinear systems with strict-feedback, the celebrated backstepping method is not used here. A simulation example is given to demonstrate the effectiveness of the proposed design procedure.

2.1 BACKGROUND

Nonlinear systems with strict feedback represent a wide range of nonlinear systems, and an effective control method for such systems is backstepping. The global asymptotic stabilization problem by output feedback for strict-feedback nonlinear systems is considered in [102]. In [53], the problem of almost certain finite-time stabilization for a class of stochastic strict-feedback nonlinear systems is investigated, while robust control of output feedback is considered for a class of strict-feedback time-delayed systems is considered in [44]. Although backstepping is widely used in controlling nonlinear systems with strict feedback, the huge increase in computational complexity is unacceptable when the system dimension is high.

The increasing gain control method has been effectively applied to strict-feedback nonlinear systems. The problem of global state regulation by output feedback is investigated in [60], where a universal output feedback controller is given to ensure that all closed-loop system signals are bounded. Based on the finite-time Lyapunov stability theorem, [146] presents a finite-time control strategy for strict-feedback nonlinear systems with both low-order and high-order nonlinearities. The stabilization problem of time-delayed nonlinear systems is considered in [154] where a non-smooth dynamic state compensator is constructed. The adaptive control was considered in [130] for single input uncertain nonlinear systems with input saturation and unknown external disturbance. The problem of even-trigger-based adaptive control was considered in [132] for a class of uncertain nonlinear systems.

It is worth noting that all the above results apply only to time-invariant systems. How to extend the above results to time-varying systems is a non-trivial task [13].

Robust adaptive control of strict-feedback time-varying nonlinear systems with unknown control directions is investigated in [46], and a Nussbaum-type function is developed such that all closed-loop states are bounded. The global finite-time stabilization problem for time-varying nonlinear systems is studied [86], and a control strategy combining backstepping and finite-time control theory is proposed. Based on the dynamic gain control method, the finite-time stabilization problem is also considered in [148] for time-varying nonlinear systems. Despite the above progress, results for controlling complex time-varying nonlinear systems are still limited.

In this chapter, the problem of stabilization of time-varying nonlinear systems with the strict-feedback form is studied. The main feature of this chapter can be summarized as follows.

- The nonlinear functions considered in this chapter are not only time-varying, but also depend on the system output, leading to a more generalized system model. The usual controllers with bounded gain cannot effectively handle the time-varying function of the concerned system. For this reason, two new gain functions are introduced in this chapter.
- A method of feedback control with an increasing gain is proposed. A state feedback controller with a dynamic gain and a time-varying gain is designed, where the dynamic gain is used to cope with the nonlinearities of the system and the time-varying gain is constructed to cope with the time-varying function. By introducing a time-varying gain and a new dynamic gain, the system states converge to zero.

2.2   PROBLEM DESCRIPTION

Consider the strict-feedback nonlinear system

$$
\begin{cases}
\dot{x}_1 = x_2 + f_1(x_1), \\
\dot{x}_2 = x_3 + f_2(x_1, x_2), \\
\quad\vdots \\
\dot{x}_n = u + f_n(x_1, x_2, \ldots, x_n),
\end{cases}
\tag{2.1}
$$

where $x = (x_1, x_2, \ldots, x_n)^T \in \mathbb{R}^n$ is the system state, and $u \in \mathbb{R}$ is the control input. The initial time instant is denoted as t_0, and the initial state is $x(t_0)$.

Functions $f_1, f_2, \ldots, f_{n-1}$ are continuous and satisfy the following two assumptions.

Assumption 2.1. *For any $x_1, x_2, \ldots, x_n, u \in \mathbb{R}$, it holds that*

$$
|f_i(x_1, \ldots, x_i)| \le \phi_1(t)\phi_2(x_1)(|x_1| + |x_2| + \ldots + |x_i|), \ i = 1, 2, \ldots, n,
$$

where $\phi_1(t) \ge 1$ is a non-decreasing smooth function, and $\phi_2(x_1)$ is a continuous function.

Assumption 2.2. *The time-varying function $\phi_1(t)$ cannot escape in finite time, and there is a positive constant ε such that*

$$\sup_{t\in[t_0,+\infty)} \frac{\dot{\phi}_1(t)}{\phi_1^2(t)} \le \varepsilon.$$

Under these two assumptions, the problem to be addressed is to design a controller such that system (2.1) is globally asymptotically stable.

Remark 2.1. *Even though system (2.1) under Assumption 2.1 is in the strict-feedback form, the nonlinear functions are complex as their growth rate depends on not only the system states, but also the time-varying function $\phi_1(t)$. How to determine the control gain to dominate $\phi_1(t)$ and $\phi_2(x_1)$ is a non-trivial task.*

Remark 2.2. *Assumption 2.2 is an imposed condition on the nonlinear term $\phi_1(t)$. This condition includes the following cases:*

- *If $\phi_1(t)$ has a limit, i.e., $\lim_{t\to+\infty} \phi_1(t) = c$ for a constant $c \in (0,\infty)$, then it is bounded and satisfies Assumption 2.2.*
- *For any positive constants c_1, c_2, the function $\phi_1(t) = c_1 e^{c_2 t}$ satisfies Assumption 2.2.*

2.3 CONTROL DESIGN

Consider the state transformation

$$z_1 = x_1, \; z_2 = x_2 + \phi_1(t)M(x_1)x_1, \; z_3 = x_3, \; \ldots, z_n = x_n,$$

where $M(x_1)$ is a smooth function with respect to the variable x_1. Then, we can get

$$\begin{cases} \dot{z}_1 = -\phi_1(t)M(z_1)z_1 + z_2 + f_1(x_1), \\ \dot{z}_2 = z_3 + f_2(x_1,x_2) + \dfrac{d\phi_1(t)}{dt}M(z_1)z_1 \\ \qquad + \phi_1(t)\dfrac{dM(z_1)z_1}{dz_1}(z_2 - \phi_1(t)M(z_1)z_1 + f_1(x_1)), \\ \dot{z}_3 = z_4 + f_3(x_1,x_2,x_3), \\ \qquad \vdots \\ \dot{z}_n = u + f_n(x_1,x_2,x_3,\ldots,x_n). \end{cases}$$

Let

$$g_1(z_1,z_2) = f_2(x_1,x_2) + \frac{d\phi_1(t)}{dt}M(z_1)z_1$$
$$+ \phi_1(t)\frac{dM(x_1)x_1}{x_1}(z_2 - \phi_1(t)M(x_1)x_1 + f_1(x_1)),$$

and

$$g_i(z_1, z_2, \ldots, z_{i+1}) = f_{i+1}(x_1, x_2, \ldots, x_i), \quad i = 2, 3, \ldots, n-1.$$

From Assumption 2.1, we can find a smooth function $\bar{\phi}_2(z_1)$ such that

$$|g_i(z_1, z_2, \ldots, z_{i+1})| \leq \bar{\phi}_1(t)\bar{\phi}_2(z_1)(|z_1| + |z_2| + \ldots + |z_{i+1}|), \quad i = 1, 2, \ldots, n-1, \tag{2.2}$$

where

$$\bar{\phi}_1(t) = (1 + \varepsilon)\phi_1^2(t) \tag{2.3}$$

and

$$(1 + \varepsilon)\phi_1^2(t) \geq \max\left\{\phi_1(t), \phi_1^2(t), \frac{d\phi_1(t)}{dt}, 1\right\}$$

is employed.

We introduce the following state transformation

$$\eta_1 = \frac{1}{h_1 h_2} z_2, \ \eta_2 = \frac{1}{h_1^2 h_2^2} z_3, \ \ldots, \ \eta_{n-1} = \frac{1}{h_1^{n-1} h_2^{n-1}} z_n,$$

where $h_1 \geq 1$ is a dynamic gain, and $h_2 \geq 1$ is a time-varying gain. The time-varying control input u is designed as

$$u = -h_1^n h_2^n (k_1 \eta_1 + k_2 \eta_2 + \ldots + k_{n-1} \eta_{n-1}), \tag{2.4}$$

where k_1 to k_{n-1} are constants to be determined.

Let $\eta = (\eta_1, \eta_2, \ldots, \eta_{n-1})^T$, then the closed-loop system is rewritten as

$$\dot{\eta} = h_1 h_2 \tilde{A} \eta + G - \frac{\dot{h}_1}{h_1} D\eta - \frac{\dot{h}_2}{h_2} D\eta, \tag{2.5}$$

where

$$\tilde{A} = \begin{pmatrix} 0 & 1 & 0 & \cdots & 0 \\ 0 & 0 & 1 & \cdots & 0 \\ \vdots & \vdots & \vdots & & \vdots \\ 0 & 0 & 0 & \cdots & 1 \\ -k_1 & -k_2 & -k_3 & \cdots & -k_{n-1} \end{pmatrix}, \ G = \begin{pmatrix} \frac{1}{h_1 h_2} g_1(z_1, z_2) \\ \frac{1}{h_1^2 h_2^2} g_2(z_1, z_2, z_3) \\ \cdots \\ \frac{1}{h_1^{n-2} h_2^{n-2}} g_{n-2}(z_1, z_2, \ldots, z_{n-1}) \\ \frac{1}{h_1^{n-1} h_2^{n-1}} g_{n-1}(z_1, z_2, z_3, \ldots, z_n) \end{pmatrix},$$

and $D = \text{diag}\{1, 2, \ldots, n-1\}$.

If k_1 to k_{n-1} are chosen to make $\tilde{A}$ Hurwitz, then a positive definite matrix $P \in \mathbb{R}^{(n-1) \times (n-1)}$ can be found to meet

$$\tilde{A}^T P + P\tilde{A} \leq -I, \quad DP + PD - P \geq \alpha I,$$

where α is a positive constant.

2.4 STABILITY ANALYSIS

We state the main result of this chapter in the following theorem:

Theorem 2.1

Suppose that Assumptions 2.1–2.2 are satisfied. If the control gains are designed as

$$\dot{h}_1 = \frac{1}{\alpha} h_1 h_2 \max \left\{ 2(n-1) \left(\bar{\phi}_2(x_1) + \bar{\phi}_2^2(x_1) \right) \|P\| - \frac{1}{2} h_1, 0 \right\},$$

and

$$h_2(t) = \bar{\phi}_1^2(t), \quad M(x_1) \geq \phi_2(x_1) + \|P\| + 1 + 4n\varepsilon$$

where $\bar{\phi}_1(t)$ and $\bar{\phi}_2(x_1)$ are chosen according to (2.2), $h_1(t_0) \geq \max\{16n\varepsilon\lambda_{\min}(P), 1\}$ and $\lambda_{\max}(P)$ is the maximum eigenvalue of P, then system (2.1) with the control law (2.4) is globally asymptotically stable. ∎

Proof. We first analyze the stability of system (2.5). Consider the Lyapunov function candidate

$$V_1 = \eta^T P \eta$$

whose derivative is computed as

$$
\begin{aligned}
\dot{V}_1|_{(2.5)} =& \eta^T P \left(h_1 h_2 \tilde{A} \eta + G - \frac{\dot{h}_1}{h_1} D\eta - \frac{\dot{h}_2}{h_2} D\eta \right) \\
&+ \left(h_1 h_2 \tilde{A} \eta + G - \frac{\dot{h}_1}{h_1} D\eta - \frac{\dot{h}_2}{h_2} D\eta \right)^T P\eta \\
=& h_1 h_2 \eta^T \left(P\tilde{A} + \tilde{A}^T P \right) \eta - \frac{\dot{h}_1}{h_1} \eta^T (PD + DP) \eta \\
&- \frac{\dot{h}_2}{h_2} \eta^T (PD + DP) \eta + 2\eta^T PG \\
\leq& - h_1 h_2 \|\eta\|^2 - \alpha \frac{\dot{h}_1}{h_1} \|\eta\|^2 - \left(\frac{\dot{h}_1}{h_1} + \frac{\dot{h}_2}{h_2} \right) V_1 + 2\eta^T PG.
\end{aligned}
\tag{2.6}
$$

For the nonlinear term G, it holds that

$$
\begin{aligned}
\left| \frac{1}{h_1^i h_2^i} g_i(z_1, z_2, \ldots, z_{i+1}) \right| \leq& \frac{1}{h_1^i h_2^i} \bar{\phi}_1(t) \bar{\phi}_2(z_1) \left(|z_1| + |z_2| + \ldots + |z_i| \right) \\
\leq& \bar{\phi}_1(t) \bar{\phi}_2(z_1) \left(\frac{1}{h_1^i h_2^i} |z_1| + \frac{1}{h_1^i h_2^i} |z_2| + \ldots + \frac{1}{h_1^i h_2^i} |z_i| \right) \\
\leq& \bar{\phi}_1(t) \bar{\phi}_2(z_1) \left(\frac{1}{h_1 h_2} |z_1| + |\eta_1| + \ldots + |\eta_{i-1}| \right),
\end{aligned}
$$

for $i = 1, 2, \ldots, n-1$, which indicates

$$\|G\| \leq \sqrt{n-1} \bar{\phi}_1(t) \bar{\phi}_2(z_1) \left(\frac{1}{h_1 h_2} |z_1| + \sqrt{n-1} \|\eta\| \right).$$

Back to (2.6), one can get

$$\dot{V}_1|_{(2.5)} \leq -h_1 h_2 \|\eta\|^2 - \alpha \frac{\dot{h}_1}{h_1} \|\eta\|^2 - \left(\frac{\dot{h}_1}{h_1} + \frac{\dot{h}_2}{h_2} \right) V_1$$
$$+ 2\sqrt{n-1} \bar{\phi}_1(t) \bar{\phi}_2(z_1) \|P\| \|\eta\| \left(\frac{1}{h_1 h_2} |z_1| + \sqrt{n-1} \|\eta\| \right)$$
$$\leq -h_1 h_2 \|\eta\|^2 - \alpha \frac{\dot{h}_1}{h_1} \|\eta\|^2 - \left(\frac{\dot{h}_1}{h_1} + \frac{\dot{h}_2}{h_2} \right) V_1$$
$$+ 2(n-1) \bar{\phi}_1^2(t) \left(\bar{\phi}_2(z_1) + \bar{\phi}_2^2(z_1) \right) \|P\| \|\eta\|^2 + \frac{1}{h_1^2 h_2^2} \|P\| z_1^2$$
$$\leq -\frac{h_1 h_2}{2} \|\eta\|^2 - \left(\frac{\dot{h}_1}{h_1} + \frac{\dot{h}_2}{h_2} \right) V_1 + \frac{1}{h_1 h_2} \|P\| z_1^2.$$

Let

$$V = h_1 h_2 V_1 + \frac{1}{2} z_1^2.$$

Its derivative is computed as

$$\dot{V}|_{(2.5)} \leq -\frac{h_1^2 h_2^2}{2} \|\eta\|^2 + \|P\| z_1^2 - \phi_1(t) M(z_1) z_1^2 + z_1 z_2 + z_1 f_1(x_1)$$
$$\leq -\frac{h_1^2 h_2^2}{2} \|\eta\|^2 + \|P\| z_1^2 - \phi_1(t) M(z_1) z_1^2 + h_1 h_2 z_1 \eta_1 + z_1 f_1(x_1)$$
$$\leq -\frac{h_1^2 h_2^2}{4} \|\eta\|^2 - \phi_1(t) \left(M(z_1) - 1 - \|P\| - \phi_2(z_1) \right) z_1^2.$$

It can be seen that $V(t)$ is bounded, and $V(t) \leq V(t_0), t \geq t_0$. From the design of h_1 and $M(x_1)$, they are both bounded since $\phi_2(x_1)$ and $\bar{\phi}_2(x_1)$ are bounded. Thus, since $h_2 \geq \phi_1(t)$, and $h_1 \geq 16 n \varepsilon \lambda_{\min}(P)$, we can get

$$\dot{V}|_{(2.5)} \leq -8 n \varepsilon \phi_1(t) V.$$

Next, we further consider $\omega = h_2^{2n-2} V$ whose derivative is given as

$$\dot{\omega} \leq -8 n \varepsilon \phi_1(t) \omega + (2n-2) \frac{\dot{h}_2}{h_2} \omega.$$

From (2.3), it holds that
$$h_2 = (1 + \varepsilon)^2 \phi_1^4(t),$$

and we obtain

$$\frac{\dot{h}_2}{h_2} = \frac{4 \dot{\phi}_1(t)}{\phi_1(t)} \leq 4 \varepsilon \phi_1(t),$$

where Assumption 2.1 is employed.

Then, we get

$$\dot{\omega} \leq -8\varepsilon\phi_1(t)\omega.$$

Because

$$|\phi_1(t)x_1| \leq h_2^{n-1}|x_1| \leq \sqrt{2\omega},$$

and

$$|z_i| \leq h_1^i h_2^i |\eta_i| \leq h_1^i \omega, \quad i = 2, 3, \ldots, n,$$

the convergence of $z_1, z_2, \ldots, z_n$ and $\phi_1(t)x_1$ are guaranteed. Moreover, we also have

$$x_1 = z_1, \ x_2 = z_2 - \phi_1(t)M(x_1)x_1, \ x_3 = z_3, \ \ldots, x_n = z_n,$$

and thus the convergence of states $x_1, x_2, \ldots, x_n$ is obtained. This ends the proof. $\square$

Remark 2.3. *Now we discuss the condition* $\frac{\dot{\phi}_1(t)}{\phi_1^2(t)} \leq \varepsilon$. *If* $\phi_1(t)$ *satisfies* $\frac{\dot{\phi}_1(t)}{\phi_1^2(t)} \geq \varepsilon$ *during a period* $t \in [t_1, t_2)$, *then we have*

$$\phi_1(t) \geq \frac{1}{\frac{1}{\phi_1(t_1)} - \varepsilon(t - t_1)} \geq \frac{1}{1 - \varepsilon(t - t_1)}, \quad t \in [t_1, t_2).$$

It can be seen that when $t_2 \to t_1 + \frac{1}{\varepsilon}$, $\phi_1(t_2) \to +\infty$. *Thus, this condition is to ensure that the time-varying function* $\phi_1(t)$ *cannot escape in finite time. Indeed, for the time-varying function* $c_1 e^{c_2 t}$ *with* c_1, c_2 *being any positive constants, it holds that* $\limsup_{t \to +\infty} \frac{\dot{\phi}_1(t)}{\phi_1^2(t)} = 0$.

It can be seen that the control law proposed above with an increasing gain can stabilize the system (2.1) with complex nonlinear dynamics. In fact, a control law with a constant but high gain can also stabilize system (2.1) with $\phi_1(t) = 1$. However, in order to determine the permissible value of the control gain, some information about the initial system states is needed. We present the result as follows.

Theorem 2.2

Suppose that Assumption 2.1 is satisfied with $\phi_1(t) = 1$ and $\|x(t_0)\| \leq \delta$, then system (2.1) can be semi-globally stabilized through the control

$$u = -h^{n-1}k_1 x_1 - h^{n-2}k_2 x_2 - \ldots - k_n x_n,$$

where k_1 to k_n are the constants making the polynomial $s^n + k_n s^{n-1} + \ldots + k_2 s + k_1$ Hurwitz, and h satisfies

$$h \geq \max\{4n(\theta_\delta + 1)\|P\|, 1\},$$

where $\theta_\delta = \max\{\phi_2(x_1)|\|x_1\| \leq \sqrt{\frac{\lambda_{\max}(P)}{\lambda_{\min}(P)}}\delta\}$, P is the positive definite matrix satisfying

$$\tilde{A}^T P + P\tilde{A} \leq -I$$

with

$$\tilde{A} = \begin{pmatrix} 0 & 1 & 0 & \cdots & 0 \\ 0 & 0 & 1 & \cdots & 0 \\ \vdots & \vdots & \vdots & & \vdots \\ 0 & 0 & 0 & \cdots & 1 \\ -k_1 & -k_2 & -k_3 & \cdots & -k_n \end{pmatrix}.$$

∎

Proof. Consider the state transformation

$$z_1 = x_1, \quad z_2 = \frac{1}{h}x_2, \quad \ldots, \quad z_n = \frac{1}{h^{n-1}}x_n,$$

then we can obtain

$$\dot{z} = h\tilde{A}z + F, \tag{2.7}$$

where $F = \left(f_1(x_1), \frac{1}{h}f_2(x_1,x_2), \ldots, \frac{1}{h^{n-2}}f_{n-1}(x_1,\ldots,x_{n-1}), \frac{1}{h^{n-1}}f_n(x_1,x_2,\ldots,x_n)\right)^T$.

Since h is a constant, the convergence of x is equivalent to that of z. Consider the Lyapunov function candidate

$$V = z^T Pz.$$

Its derivative is computed as

$$\begin{aligned} \dot{V}|_{(2.7)} &= \left(hz^T\tilde{A}^T + F^T\right)Pz + z^T P\left(h\tilde{A}z + F\right) \\ &= hz^T\left(\tilde{A}^T P + P\tilde{A}\right)z + 2z^T PF \\ &\leq -h\|z\|^2 + 2z^T PF. \end{aligned} \tag{2.8}$$

From Assumption 2.1, we can get

$$\begin{aligned} |\frac{1}{h^{i-1}}f_i(x_1,x_2,\ldots,x_i)| &\leq \frac{1}{h^{i-1}}\phi_2(x_1)\left(|x_1| + |x_2| + \ldots + |x_i|\right) \\ &\leq \frac{1}{h^{i-1}}\phi_2(x_1)\left(|z_1| + |hz_2| + \ldots + |h^{i-1}z_i|\right) \\ &\leq \phi_2(x_1)\left(\frac{1}{h^{i-1}}|z_1| + \frac{1}{h^{i-2}}|z_2| + \ldots + |z_i|\right) \\ &\leq \phi_2(x_1)\left(|z_1| + |z_2| + \ldots + |z_i|\right) \\ &\leq \phi_2(x_1)\sqrt{n}\|z\|, \end{aligned}$$

where $h \geq 1$ is employed. Then, the estimation of F is expressed as

$$\|F\| \leq n\phi_2(x_1)\|z\|.$$

Back to (2.8), we have

$$\dot{V}|_{(2.7)} \leq -h\|z\|^2 + 2\|P\|\|z\|\|F\| \leq -h\|z\|^2 + 2n\phi_2(x_1)\|P\|\|z\|^2.$$

Since $\|x(t_0)\| \leq \delta$, and $\phi_2(x_1(t_0)) \leq \theta_\delta$, we have $\dot{V} \leq -\frac{1}{2}h\|z\|^2$. Now, we prove the stability of the system by contradiction. Assume that t_1 is the time instant such that when $t \in [t_0, t_1)$, $\phi_2(x_1(t)) \leq \theta_\delta$, and $\phi_2(x_1(t_1)) > \theta_\delta$. Then, we have

$$\dot{V}|_{(2.7)} \leq -\frac{1}{2}h\|z\|^2, \quad t \in [t_0, t_1),$$

and

$$\lambda_{\min}(P)\|x_1(t)\|^2 \leq \lambda_{\min}(P)\|z(t)\|^2 \leq V(t)$$
$$< V(t_0) \leq \lambda_{\max}(P)\|z(t_0)\|^2 \leq \lambda_{\max}(P)\|x(t_0)\|^2,$$

where $\lambda_{\max}(P)$ and $\lambda_{\max}(P)$ are the maximum and minimum eigenvalue of P, respectively. Thus, we obtain

$$\|x_1(t)\| < \sqrt{\frac{\lambda_{\max}(P)}{\lambda_{\min}(P)}}\|x(t_0)\|.$$

Since $x(t)$ is continuous, we get

$$\|x_1(t_1)\| \leq \sqrt{\frac{\lambda_{\max}(P)}{\lambda_{\min}(P)}}\|x(t_0)\|.$$

From the definition of δ_ε, we get $\phi_2(x_1(t_1)) \leq \theta_\delta$, which contracts the assumption. This implies that

$$\dot{V} \leq -\frac{1}{2}h\|z\|^2$$

holds all the time. Since h is a positive constant, we can get that all the closed-loop signals are stable. This completes the proof. □

Therefore, choosing high gain can also stabilize the strict-feedback nonlinear system, although it gets a local result. To get a global stabilizing result, we prefer to adaptively determine the control gain. In this case, the generated closed-loop system is globally stabilized.

2.5 SIMULATION

Consider the nonlinear system

$$\begin{cases} \dot{x}_1 = x_2 + \phi(t)x_1^2, \\ \dot{x}_2 = u - \phi(t)x_1x_2, \end{cases}$$

where $x = (x_1, x_2)^T \in \mathbb{R}^2$ is the state, $u \in \mathbb{R}$ is the input, and $\phi(t)$ is a time-varying function.

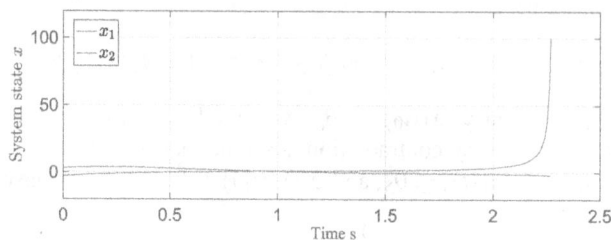

(a) $h = 2$ and $(x_1(0), x_2(0)) = (-3, 3)$.

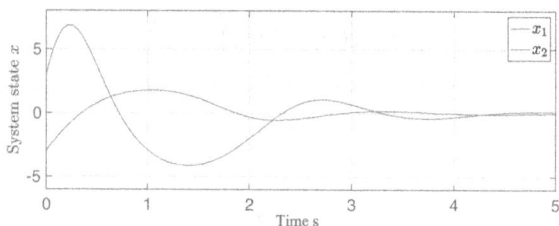

(b) $h = 10$ and $(x_1(0), x_2(0)) = (-3, 3)$.

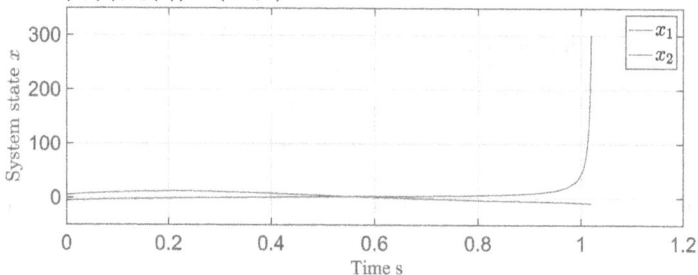

(c) $h = 10$ and $(x_1(0), x_2(0)) = (-5, 5)$.

Figure 2.1   Control performance under different gains and initial conditions.

Part I: Verification of Theorem 2.2.

In this case, we set $\phi(t) = 1$ and the initial state as $(x_1(0), x_2(0)) = (-3, 3)$. The control law is designed as

$$u = -hx_1 - 2x_2,$$

where h is a regulating parameter.

The simulation results are shown in Figure 2.1a and Figure 2.1b. From these results, it can be observed that the system states diverge when $h = 2$; while when h increases to ten, the system states become stable. However, when the initial states become $(x_1(0), x_2(0)) = (-5, 5)$, the system becomes unstable again. This illustrates that, to make the system stable, the gain h has to be carefully scheduled according to the system's initial states.

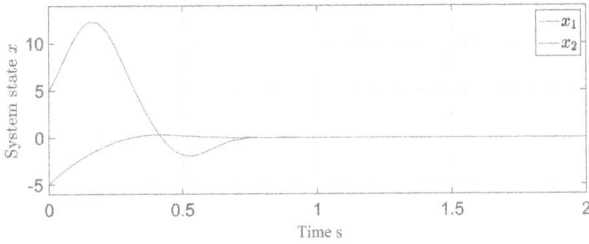

(a) The state trajectory.

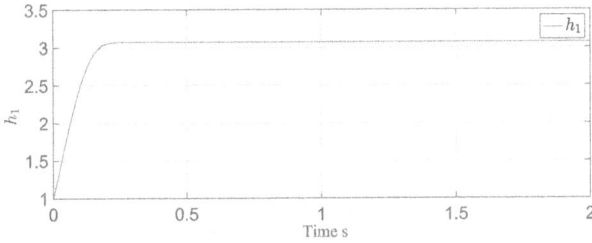

(b) The dynamic gain h_1.

Figure 2.2 Control performance with a dynamic gain and $(x_1(0), x_2(0)) = (-5, 5)$.

Part II: Verification of Theorem 2.1.

In this case, we set $\phi(t) = e^{2t}$ and design the control law as

$$u = -h_1 e^{4t} \left(x_2 + (|x_1| + 1)e^{4t} x_1\right),$$

where h_1 is the dynamic gain satisfying

$$\dot{h}_1 = \frac{1}{3} h_1 e^{2t} \max\{2x_1^2 - \frac{1}{2}h_1, 0\}, \quad h_1(0) = 1.$$

The simulation results are shown in Figure 2.2. As can be observed, h_1 initially increases but quickly becomes a constant. Even though the value of this constant (i.e., $h_1 = 3$) is smaller than the one in the above case (i.e., $h = 10$), the system can still be stabilized. This demonstrates the advantage of the control strategy with a dynamic gain over the one with a constant gain.

2.6 NOTES

In this chapter, the global state feedback control problem has been solved for non-linear systems with strict-feedback. The chapter has shown that classical high-gain feedback control or low-gain feedback control can only achieve a semi-global stabilization result for a complex nonlinear system. In order to achieve a global result and dominate the time-varying terms, the variable gain feedback control is introduced here. In contrast to the existing works, we have considered the more general cases and given the condition for time-varying terms. Under this condition and the Lyapunov stability theorem, the state-feedback stabilization or control controller was designed such that the system states converge to zero.

3 Output Feedback Control for Strict-Feedback Nonlinear Systems with Time Delays

In this chapter, constructive control techniques are proposed for the control of strict-feedback nonlinear systems with a time delay in the state. It is assumed that the uncertain nonlinearities are bounded by functions of the output multiplied by unmeasured states or delayed states. Based on the use of a linear dynamic high-gain observer, a delay-independent output feedback controller is explicitly constructed to make the closed-loop system asymptotically stable. A simulation example is used to demonstrate the effectiveness of the proposed control strategy.

3.1 BACKGROUND

In recent decades, the problem of global stabilization of triangular structural nonlinear systems by output feedback control has received considerable attention. As investigated in [93], for the global stabilization of nonlinear systems by output feedback, some additional growth conditions for immeasurable states of the system are usually required. In [106], under a linear growth assumption, a class of uncertain nonlinear systems were considered by output feedback control. In [102], the problem of robust output feedback control is considered for systems in lower triangular form, and a global Lipschitz-like condition on immeasurable states with an output dependent incremental rate is used.

The existence of time delays can lead to instability. Therefore, the stability analysis of systems with time delays becomes one of the main concerns of researchers [111, 38, 12]. The Lyapunov-Krasovskii method and the Lyapunov-Razumikhin method are usually used in the stability analysis of linear time-delayed systems, and the main conclusions are often given in the form of linear matrix inequalities. However, compared to linear systems, the results for nonlinear time-delayed systems are still limited. Typical constructive control methods for nonlinear systems include backstepping, forwarding and saturation design [59, 57].

The dynamic gain control approach has recently been applied to the stabilization control of nonlinear systems in triangular form. It has been shown that this control approach plays an important role in coping with system nonlinearities [103, 147]. However, there are still limited works that use this approach to solve the problem of global stabilization of nonlinear systems with output feedback in the form of a lower triangle, which motivates us to consider this interesting and challenging problem.

The contributions of this chapter can be characterized by the following features:

- The dynamic gain control approach is used to study the global output feedback stabilization problem of lower triangular nonlinear time-delay systems. The usual backstepping method is not used here, avoiding the problem of "explosion of complexity" caused by the repeated differentiation of virtual controllers.
- The bounding functions of the interaction terms are allowed to depend on unmeasured states, hence the result obtained is more general than the one in [11].

3.2   PROBLEM DESCRIPTION

In this chapter, the nonlinear time-delay system is modeled as

$$\begin{aligned}
\dot{x}_i &= x_{i+1} + f_i(t,x,x(t-d),u), \ i = 1,2,\ldots,n-1, \\
\dot{x}_n &= u + f_n(t,x,x(t-d),u), \\
y &= x_1,
\end{aligned} \tag{3.1}$$

where $x = (x_1,x_2,\ldots,x_n)^T \in \mathbb{R}^n$ and $y \in \mathbb{R}$ represent the system state and output, respectively; $f_i : \mathbb{R}^{2n+2} \to \mathbb{R}$, $i = 1,2,\ldots,n$, are continuous functions, $d \geq 0$ denotes the time delay.

The control objective here is to design a controller with output feedback that can stabilize system (1) in the presence of time delays. For the sake of controller design, the following assumption and lemma are given as follows:

Assumption 3.1. *There exists a non-negative constant C and a smooth function $L(y)$ such that the following inequalities hold:*

$$\begin{aligned}
\left| f_1(t,x,x(t-d),u) \right| &\leq C\big(|x_1| + |x_1(t-d)|\big), \\
\left| f_i(t,x,x(t-d),u) \right| &\leq L(y)\sum_{j=1}^{i}\big(|x_j| + |x_i(t-d)|\big).
\end{aligned} \tag{3.2}$$

Lemma 3.1: [103]

There exists a positive constant τ, and real numbers $a_k, b_k, k = 2,3,\ldots,n$, and positive definite matrices P, Q satisfying the following inequalities

$$\begin{aligned}
PA + A^T P &\leq -I_{n-1}, & PD + DP - P &\geq \tau I, \\
QB + B^T Q &\leq -I_{n-1}, & QD + DQ - Q &\geq \tau I,
\end{aligned}$$

where I_{n-1} is an identity matrix with a dimension $n - 1$, $A = G - H_a K_a$, $K_a = (1,0,\ldots,0) \in \mathbb{R}^{1\times(n-1)}$, $H_a = (a_2,a_3,\ldots,a_n)^T \in \mathbb{R}^{(n-1)\times 1}$, $B = G - K_b H_b$,

$K_b = (0,0,\ldots,1)^T \in \mathbb{R}^{(n-1)\times 1}$, $H_b = (b_2,b_3,\ldots,b_n) \in \mathbb{R}^{1\times(n-1)}$, $D = \mathrm{diag}(1,2,\ldots,n-1)$, $G = \begin{pmatrix} \mathbf{0}^T_{1\times(n-2)} & I_{n-2} \\ 0 & \mathbf{0}_{1\times(n-2)} \end{pmatrix}$. ■

3.3 CONTROL DESIGN

In this section, we will first design a reduced-order observer, based on which the output feedback controller will be given accordingly.

A reduced-order observer for system (3.1) is constructed as follows:

$$\dot{\hat{x}}_i = \hat{x}_{i+1} + a_{i+1}\rho^i y - (i-1)a_i\dot{\rho}\rho^{i-2}y - a_i\rho^{i-1}(\hat{x}_2 + a_2\rho y), \quad i = 2,3,\ldots,n-1,$$
$$\dot{\hat{x}}_n = u - (n-1)a_n\dot{\rho}\rho^{n-2}y - a_n\rho^{n-1}(\hat{x}_2 + a_2\rho y),$$

$$(3.3)$$

where a_i, $i = 2,3,\ldots,n$, are constants given by Lemma 3.1, and ρ is the state of the following system:

$$\dot{\rho} = \tfrac{\rho}{\tau}\max\{-\tfrac{\rho}{4} + \upsilon(y),0\},$$
$$\rho(t) = \rho_0 \geq 1, \text{ for } t \in [-d,0],$$

$$(3.4)$$

with τ being a constant given by Lemma 3.1, and $\upsilon(y)$ is a continuously differentiable non-negative function to be specified later.

Define $\eta_i = \hat{x}_i + a_i\rho^{i-1}y - x_i$, $i = 2,3,\ldots,n$. Then we have

$$\dot{\eta}_i = \eta_{i+1} - a_i\rho^{i-1}\eta_2 + a_i\rho^{i-1}f_1 - f_i,$$
$$\dot{\eta}_n = -a_n\rho^{n-1}\eta_2 + a_n\rho^{n-1}f_1 - f_n.$$

$$(3.5)$$

The following scaled states can be introduced

$$\xi_i = \frac{\eta_i}{\rho^{i-1}}, \quad i = 2,3,\ldots,n.$$

$$(3.6)$$

By (3.5) and (3.6), the following matrix form is obtained

$$\dot{\xi} = \rho A\xi - \frac{\dot{\rho}}{\rho}D\xi + \Psi,$$

$$(3.7)$$

where $\xi = (\xi_2,\xi_3,\ldots,\xi_n)^T$ and $\Psi = \left(a_2 f_1 - \tfrac{f_2}{\rho}, a_3 f_1 - \tfrac{f_3}{\rho^2},\ldots,a_n f_1 - \tfrac{f_n}{\rho^{n-1}}\right)^T$.

Remark 3.1. *For $t \geq 0$, it is obvious that ρ in (3.5) has the following properties: $\dot{\rho} \geq 0$, $\rho \geq 1$, $\tfrac{\rho}{4} + \tfrac{\dot{\rho}}{\rho}\tau \geq \upsilon(y)$, and $\rho(t) \geq \rho(t-d) \geq 1$. Therefore, ρ is a monotonically non-decreasing function. If the functions $\upsilon_1(y)$ and $\upsilon_2(y)$ are bounded, there exists a finite time T such that $\dot{\rho}(t) = 0$, $\forall t \in [T,+\infty)$.*

Define the following states as

$$z_2 = \frac{\hat{x}_2 + My}{\rho} + a_2 y,$$
$$z_i = \frac{\hat{x}_i}{\rho^{i-1}} + a_i y, \quad i = 3,4,\ldots,n.$$

$$(3.8)$$

where M is a positive constant to be specified later.

Now, the output feedback control strategy is designed as

$$u = -\rho^n(b_2 z_2 + b_3 z_3 + \ldots + b_n z_n). \tag{3.9}$$

The time derivative of (3.8) can be rewritten compactly as

$$\dot{z} = \rho B z - \frac{\dot{\rho}}{\rho} D z - \rho H_a K_a \xi + H_a f_1 + K_a^T \Upsilon, \tag{3.10}$$

where $z = (z_2, z_3, \ldots, z_n)^T$, $\Upsilon = M\left(\frac{-My}{\rho} + K_a z - K_a \xi + \frac{f_1}{\rho}\right)$.
In addition, it can be seen from (3.1), (3.8) and (3.10) that

$$\dot{y} = -My + \rho K_a z - \rho K_a \xi + f_1. \tag{3.11}$$

3.4 STABILITY ANALYSIS

Based on the observer and controller designed above, we have the following theorem:

Theorem 3.1

For controller (3.9) under Assumption 3.1, if $M(y)$ satisfies (3.17), $\upsilon_1(y)$ and $\upsilon_2(y)$ are determined by (3.18), and the constants τ, a_i, b_i, $i = 2, 3, \ldots, n$, are determined by Lemma 3.1, then the closed-loop system consisting of (3.1), (3.2), (3.3), (3.4), (3.8) and (3.9) is globally asymptotically stable. ∎

Proof. Choose a Lyapunov function candidate

$$V_y = \frac{1}{2} y^2. \tag{3.12}$$

From (3.11) and (3.12), one can get

$$\dot{V}_y|_{(3.11)} = -My^2 + \rho y K_a z - \rho y K_a \xi + y f_1.$$

Then, we have

$$\rho y K_a z - \rho y K_a \xi + y f_{1,j} \leq \frac{\rho^2}{4} \|z\|^2 + C_1 \rho^2 \|\xi\|^2 + C_2 y^2 + C_3 y^2(t - d), \tag{3.13}$$

where C_i are constants independent of M.
Consider another Lyapunov function

$$V_\xi = \rho \xi^T P \xi,$$

whose derivative can be given by

$$\dot{V}_\xi|_{(3.7)} \leq -\rho^2 \|\xi\|^2 - \tau \dot{\rho} \|\xi\|^2 + 2\rho \Psi^T P \xi.$$

Using Assumption 3.1, (3.6), and (3.8), one can get

$$\rho\left|\frac{f_i}{\rho^{i-1}}\right| \leq L(y)\Big((1+M)|y| + r\sum_{j=2}^{i}(|\xi_i| + |z_i|) + (1+M)|y(t-d)| \\ + r^{\frac{1}{2}}r^{\frac{1}{2}}(t-d)\sum_{j=2}^{i}(|\xi_i(t-d)| + |z_i(t-d)|)\Big), \; i = 2,3,\ldots,n.$$

By Young's inequality, there exist a known positive constant c_1, a smooth function $\varphi_1(y)$ independent of $M(y)$ and ρ, and a continuous function $\varphi_2(y)$, such that

$$2\rho\Psi^T P\xi \leq \frac{\rho^2}{2}\|\xi\|^2 + C_4 y^2 + C_4 y^2(t-d) + \rho F_1(y,M)\|\xi\|^2 + \rho F_2(y)\|z\|^2 \\ + k_1\rho(t-d)\|\xi(t-d)\|^2 + k_2\rho(t-d)\|z(t-d)\|^2 \\ + C_5 y^2 + C_5 y^2(t-d),$$
(3.14)

where $C_j > 0$, $j = 4,5$, are constants independent of M, F_1 is a function of y and M, F_2 is a function of y, k_1 and k_2 are positive constants.

Now consider the Lyapunov function candidate

$$V_z = \rho z^T Q z.$$

From (3.10) and (3.14), we can get

$$\dot{V}_z|_{(3.10)} \leq -\rho\|z\|^2 - \tau\dot{\rho}\|z\|^2 + 2|\rho^2 z^T Q H_a K_a \xi| + 2|\rho z^T Q H_a f_1| + 2|\rho z^T Q K_a^T \Upsilon|.$$

Similar to the argument for (3.14), there exist a smooth function $\varphi_3(y)$ independent of $M(y)$ and ρ, and the continuous functions $\varphi_i(y)$, $i = 4,5,6$, such that

$$2|\rho^2 z^T Q H_a K_a \xi| \leq \frac{\rho^2}{4}\|z\|^2 + C_6\rho^2\|\xi\|^2,$$

$$2|\rho z^T Q H_a f_1| \leq \frac{\rho^2}{4}\|z\|^2 + C_7 y^2 + C_7 y^2(t-d),$$
(3.15)

$$2|\rho z^T Q K_a^T \Upsilon| \leq k_3(M)\rho\|z\|^2 + k_4(M)\rho\|\xi\|^2 + C_8 y^2 + C_9 y^2(t-d),$$

where $C_j > 0$, $j = 6,7,8,9$, are constants independent of M, k_3 and k_4 are functions of M.

We consider the following Lyapunov function candidate for the whole system

$$V = \tilde{V}_y + \tilde{V}_z + 4(C_1 + C_6)\tilde{V}_\xi,$$

where

$$\tilde{V}_y = V_y + \int_{t-d}^{t} C_3 y^2(s)ds,$$

$$\tilde{V}_z = V_\xi + \int_{t-d}^{t} (k_1\rho(s)\|\xi(s)\|^2 + k_2\rho(s)\|z(s)\|^2 + (C_4 + C_5)y^2(s))ds,$$

$$\tilde{V}_\xi = V_z + \int_{t-d}^{t} (C_7 + C_9)y^2(s)ds.$$

It can be calculated by (3.14), (3.15) and (3.13) that

$$\dot{V}|_{(3.7)(3.10)(3.11)} \leq \left(-M+C_2+C_3+8(C_1+C_6)(C_4+C_5)+2C_7+C_8+C_9\right)y^2$$
$$+\rho\left(-\frac{\rho}{4}-\tau\frac{\dot{\rho}}{\rho}+4(C_1+C_6)(F_2+k_2)+k_3(M)\right)\|z\|^2$$
$$+4\rho(C_1+C_6)\left(-\frac{\rho}{4}-\tau\frac{\dot{\rho}}{\rho}+F_1(y,M)+\frac{k_4(M)}{4(C_1+C_6)}+k_1\right)\|\xi\|^2.$$

$$(3.16)$$

Now, M can be specified as

$$M > \delta_1 + C_2 + C_3 + 8(C_1+C_6)(C_4+C_5)+2C_7+C_8+C_9, \qquad (3.17)$$

where δ_1 can be any positive constant. Then, we can find a smooth function $\upsilon(y)$ such that the following inequities hold for any s:

$$\begin{aligned}\upsilon(s) &\geq \delta_2 + F_1(s,M) + \frac{k_4(M)}{4(C_1+C_6)}+k_1,\\ \upsilon(s) &\geq \delta_3 + 4(C_1+C_6)(F_2(s)+k_2)+k_3(M),\end{aligned} \qquad (3.18)$$

where δ_2 and δ_3 are positive constants.

Based on (3.4), (3.17) and (3.18), (3.16) can be rewritten as

$$\dot{V}|_{(3.7)(3.10)(3.11)} \leq -\delta_1 y^2 - 4\rho(C_1+C_6)\delta_2\|\xi\|^2 - \rho\delta_3\|z\|^2. \qquad (3.19)$$

Next, we shall prove that all states of the closed-loop system are bounded. To this end, defining

$$\bar{V} = \int_0^V S(\tau)d\tau,$$

where S is a non-decreasing function satisfying $S(s) > 0$, $\forall s \geq 0$, we have

$$\begin{aligned}\dot{\bar{V}}|_{(3.7)(3.10)(3.11)} &\leq S(V)\dot{V}\\ &\leq S(V)(-\delta_1 y^2 - 4\rho(C_1+C_6)\delta_2\|\xi\|^2 - \rho\delta_3\|z\|^2)\\ &\leq S(V)(-\delta_1 y^2)\\ &\leq -\delta_1 S(\tfrac{y^2}{2})y^2.\end{aligned} \qquad (3.20)$$

On the other hand, let $\rho^* = 4\upsilon(0)$ and

$$V_r(r) = \tau[\rho - \rho^* - \rho^* \ln(\tfrac{\rho}{\rho^*})],$$

which is continuously differentiable, proper, and non-negative in $(0, +\infty)$. As in [14], we can prove that

$$\dot{V}_r(r) \leq [\upsilon(y) - \upsilon(0)]^2. \qquad (3.21)$$

From (3.20) and (3.21), we obtain that

$$\frac{d\{\bar{V}+V_r(r)\}}{dt} \leq -\delta_1 S(\tfrac{y^2}{2})y^2 + [\upsilon(y) - \upsilon(0)]^2 \leq 0$$

by appropriate choice of S. As a result, $\rho(t)$ is bounded, hence all states of the closed-loop system (3.4), (3.7), (3.10), and (3.11) are bounded. From (3.19) and the definition of V, one can conclude that y is bounded, which implies boundedness of $\upsilon(y)$.

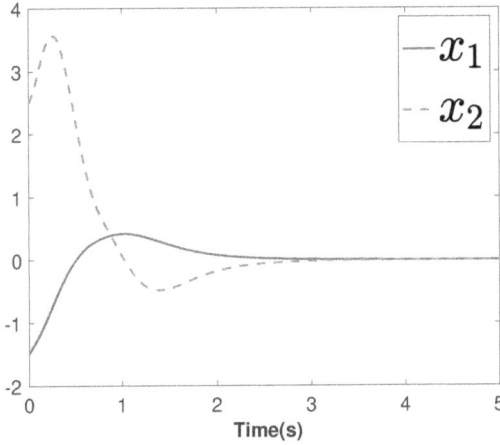

Figure 3.1 Trajectories of x_1 and x_2.

By system (3.4), one has $\dot{\rho}(t) = 0$, if $\rho \geq \max\{4v(y), 1\}$. Then, the boundedness of $v(y)$ implies the boundedness of ρ. It follows from (3.19) and the boundedness of ρ that all the signals of systems (3.7), (3.10) and (3.11) converge to zero. Therefore, based on the definitions (3.6) and (3.8) and the boundedness of ρ, one concludes that the closed-loop system is globally asymptotically stable. The proof of Theorem 3.1 is completed here.   □

3.5 SIMULATION

Consider a nonlinear time-delay system of the form:

$$\begin{aligned}
\dot{x}_1 &= x_2 + \tfrac{1}{10}\ln(1+x_1^2(t-0.5)), \\
\dot{x}_2 &= u + \tfrac{1}{6}e^{x_1}x_2^2(t-0.5), \\
y &= x_1.
\end{aligned} \tag{3.22}$$

Based on the proof of Theorem 3.1, a global output feedback controller can be designed as

$$u = -\tfrac{\rho^2}{2}\left(\tfrac{\hat{x}_2 + 1.7781y}{\rho} + \tfrac{1}{2}y\right),$$

where the state observer is designed as

$$\dot{\hat{x}}_2 = -\rho\hat{x}_2 - 0.8891\rho y - \tfrac{1}{2}\rho^2 y - \tfrac{1}{2}\dot{\rho}y,$$

and the dynamic gain is designed as

$$\dot{\rho} = \rho \max\left\{20e^y + 30 - \tfrac{\rho^2}{4}, 0\right\}, \tag{3.23}$$

with $\rho(s) = 1$, for $s \in [-0.5, 0]$.

The simulation results are shown in Figures 3.1–3.2. Figure 3.1 shows the state response of the closed-loop control system consisting of (3.22)–(3.23), with initial

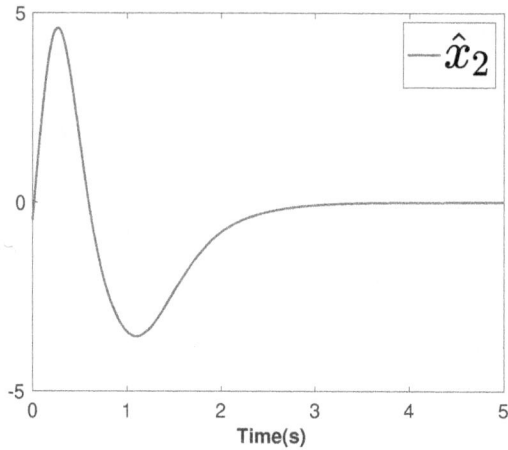

Figure 3.2   Trajectory of $\hat{x}_2$.

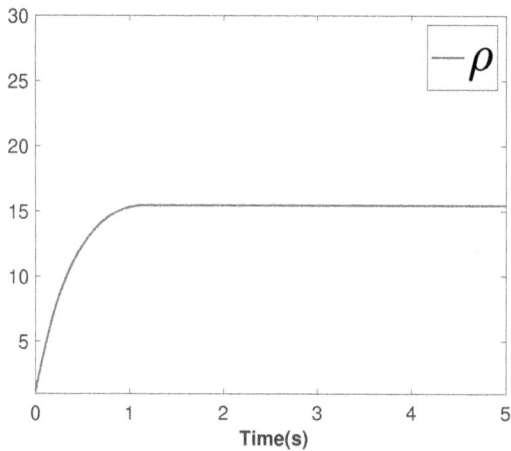

Figure 3.3   Trajectory of ρ.

condition $(x_1(s), x_2(s), \hat{x}_2(s)) = (-1.5, 2.5, -0.6), s \in [-0.5, 0]$. It can be observed from Figure 3.1 that asymptotic stabilization has been achieved.

3.6   NOTES

In this chapter, we have studied the output feedback control problem for a class of nonlinear time-delayed systems in lower triangular form. In general, it is a fact that the gains of conventional stabilizing controllers for nonlinear systems are very high, which can make the transient behaviors unsatisfactory. In contrast, the dynamic gain constructed in this chapter is monotonically increasing and converges to a constant with a reasonable amplitude. Therefore, the proposed dynamic gain approach can effectively improve the transient response of the system. Moreover, unlike most existing approaches, the design procedure does not require the well-known backstepping procedure.

4 Fixed-Time Control for Strict-Feedback Nonlinear Systems

In this chapter, a fixed-time stabilization (FSC) method for a class of nonlinear systems with strict feedback and unmodeled system dynamics is presented. Specifically, a set of auxiliary variables is first introduced by a state transformation. These variables combine the original system states and the two introduced dynamic parameters, facilitating the closed-loop system stability analysis. Then the two dynamic parameters are carefully designed using the Lyapunov method to ensure that all closed-loop system states are globally fixed-time stable. The two designed dynamic parameters depend on the system states and not on a time-varying function, so that the proposed controller is still valid beyond the given fixed-time convergence instant. The effectiveness of the proposed method is demonstrated by a simulation example.

4.1 BACKGROUND

Fixed-time stabilizing controls (FSCs) have been studied extensively in the last decade. Before a setting time, FSCs can ensure that the trajectories of the system converge to zero for any initial conditions. This kind of control was firstly formulated by [101] to solve a stabilization problem for uncertain linear plants. It can provide rapid response and high precision properties, which is particularly useful in some scenarios. For example, based on the high tracking precision, [100] developed the fixed-time consensus protocols for wheeled mobile robots. Also, FSC offers advantages in the presence of uncertain disturbances and inherent nonlinear dynamics, which is desirable in mechanical and electromechanical applications [16, 18, 40].

The design of FSCs can be divided into two categories: time-dependent control and state-dependent control. The time-dependent control is also called preset time control [117, 99], where a time-varying function is used to regulate the rate of convergence. This function would converge to zero or infinity at the given time, making the rate of convergence infinite. Although this type of control can cause the system states to converge to zero at any time, the control becomes invalid after the specified settling time. To deal with this problem, a switching strategy has to be adopted which complicates the system control design [128, 8].

For the state-dependent control design, various technologies have been introduced in the existing literature. A hybrid control algorithm was introduced by combining a finite-time stabilizing control and a fixed-time attracting control in [101]. To avoid the chattering regimes from hybrid controls, a non-hybrid control strategy with an involution operation sign was developed for both linear and nonlinear systems in

[65]. Through proposing a condition on a state-depended function, [45] proposed a continuous FSC method for nonlinear systems. The implicit Lyapunov functions were introduced in [87] to construct fixed-time observers. Recently, [121] studied the fixed-time fuzzy tracking control problem for a class of unknown nonlinear systems. The fixed-time stabilization problem for linear systems with input delay was also studied in [159]. It should be noted, however, that the control structure of the above result has complicated forms that are not easy to implement.

The dynamic gain control approach is widely used to solve the stabilization problems of nonlinear systems. This approach is able to deal with system uncertainties while guaranteeing the desired control performance. With the help of the dynamic gain control approach, the stabilization problems were solved in [103, 148, 80]. It is shown in [103, 148, 80] that the control designed by the dynamic gains has a simple linear form which can greatly simplify the controller design. This chapter aims to design an FSC control method for uncertain nonlinear systems by using the dynamic gain control approach. The main features of the proposed method include the following:

- The proposed controller consists of two dynamic parameters and has a simple quasi-linear form. The two dynamic parameters are delicately designed using the Lyapunov method to ensure that all closed-loop system states are stable in time. Compared to existing results, the "explosion of complexity" problem of backstepping control is successfully avoided.
- Different from the pre-specified finite time controllers proposed in [117, 99], the dynamic parameters in our designed controller depend on the system states and not on a time-varying function. Therefore, the control system using our proposed controller can continue to operate beyond the given fixed-time instant without requiring any change in the control strategy. This is particularly useful in the case where the given point in time is not precisely determined.

4.2   PROBLEM DESCRIPTION

4.2.1   FIXED-TIME STABILITY

Consider

$$\dot{x}(t) = g(x(t)), \tag{4.1}$$

where $x(t) \in \mathbb{R}^n$ is the system state. Function $g(\cdot) : \mathbb{R}^n \to \mathbb{R}^n$ is continuous and satisfies $g(0) = 0$. The initial time instant is assumed to be 0 and the initial state is denoted as $x_0 \in \mathbb{R}^n$.

Definition 4.1 ([3, 47], Globally Finite-time Stable). *For the system described by (4.1), it is globally finite-time stable at equilibrium $x = 0$ if it is Lyapunov stable and finite-time attractive, i.e., there exists a local bounded function $T : \mathbb{R}^n \to \mathbb{R}_+ \cup \{0\}$ such that $x(t;x_0) = 0$ for all $t \geq T(x_0)$, where $x(t;x_0)$ is a solution of (4.1) with initial state $x_0 \in \mathbb{R}^n$. The function $T(x_0)$ is called the settling time function.*

Definition 4.2 ([101], Globally Fixed-time Stable). *It is said that system (4.1) is globally fixed-time stable at equilibrium $x = 0$ if it is globally finite-time stable and the settling time function $T(x_0)$ is globally bounded by some positive constant $T_{\max} > 0$, i.e., $T(x_0) \leq T_{\max}, \forall x_0 \in \mathbb{R}^n$. The constant $T_{\max}$ is called the setting time.*

4.2.2 PROBLEM DESCRIPTION

Consider the nonlinear system

$$\begin{cases} \dot{x}_1 = x_2 + f_1(x_1), \\ \dot{x}_2 = x_3 + f_2(x_1, x_2), \\ \quad\vdots \\ \dot{x}_n = u + f_n(x_1, x_2, \ldots, x_n), \end{cases} \quad (4.2)$$

where $x = (x_1, x_2, \ldots, x_n)^T \in \mathbb{R}^n$ is the system state, $u \in \mathbb{R}$ is the system input, and $f_i(\cdot), i = 1, 2, \ldots, n$, are continuous functions satisfying the following assumption.

Assumption 4.1. *For any $x_1, x_2, \ldots, x_n \in \mathbb{R}$, it holds that*

$$|f_i(x_1, x_2, \ldots, x_i)| \leq c(|x_1| + |x_2| + \ldots + |x_i|), \quad i = 1, 2, \ldots, n,$$

where c is a positive constant.

The problem to be addressed is to design a fixed-time control strategy for (4.2) via the variable control gain technique. To this end, we present the following two lemmas.

Lemma 4.1: Jensen's Inequality

Let $0 < \tau < 1$, $a, b \geq 0$. Then it holds that

$$(a + b)^\tau \leq a^\tau + b^\tau.$$

∎

Lemma 4.2: [148]

Let a, b, γ be positive real numbers, and $\tau \in (0, 1)$. Then, we have

$$-\frac{a^2}{b^{1-\tau}} \leq -\frac{1}{1+\tau}\gamma^{-1}a^{1+\tau} + \frac{1-\tau}{1+\tau}\gamma^{-\frac{2}{1-\tau}}b^{1+\tau}.$$

∎

4.3 CONTROL DESIGN

System (4.2) can be expressed in the matrix form

$$\dot{x} = Ax + Bu + F(x),$$

where

$$A = \begin{pmatrix} 0 & 1 & 0 & \cdots & 0 \\ 0 & 0 & 1 & \cdots & 0 \\ \vdots & \vdots & \vdots & & \vdots \\ 0 & 0 & 0 & \cdots & 1 \\ 0 & 0 & 0 & \cdots & 0 \end{pmatrix}, \quad B = \begin{pmatrix} 0 \\ 0 \\ \vdots \\ 0 \\ 1 \end{pmatrix}, \quad F = \begin{pmatrix} f_1(x_1) \\ f_2(x_1,x_2) \\ \cdots \\ f_{n-1}(x_1,\ldots,x_{n-1}) \\ f_n(x_1,x_2,\ldots,x_n) \end{pmatrix}.$$

Let

$$D = \operatorname{diag}\{1,2,\ldots,n-1,n\}.$$

Then, we can find a vector $K = (k_1, k_2, \ldots, k_n) \in \mathbb{R}^n$ and a positive definite matrix $P \in \mathbb{R}^{n \times n}$ such that

$$\begin{aligned} (A - BK)^T P + P(A - BK) &\leq -I, \\ 0 \leq PD + DP &\leq \beta_1 I, \end{aligned} \tag{4.3}$$

where β_1 is a positive constant.

Then, a constant $\tau \in (0,1)$ can be found to meet

$$0 \leq (2n+2)P - (1-\tau)(PD+DP) \leq \beta_2 I, \tag{4.4}$$

with β_2 being a positive constant.

The control input u is designed as

$$u = -k_1 \frac{r_2^n}{r_1^{n(1-\tau)}} x_1 - k_2 \frac{r_2^{n-1}}{r_1^{(n-1)(1-\tau)}} x_2 - \ldots - k_n \frac{r_2}{r_1^{1-\tau}} x_n, \tag{4.5}$$

where r_1, r_2 are dynamic gains to be designed.

Now the control design problem is reduced to the design of r_1, r_2. We introduce the new variables

$$z_i = \frac{x_i}{r_1^{n+1-i+i\tau} r_2^i}, \quad i = 1, 2, \ldots, n. \tag{4.6}$$

Let $z = (z_1, z_2, \ldots, z_n)^T \in \mathbb{R}^n$ and denote α_1, α_2 as the minimum and maximum eigenvalues of P. Then, the dynamics of r_1 and r_2 can be designed as

$$\begin{aligned} \dot{r}_1 &= -\frac{1}{4\beta_2} r_1^\tau r_2 + \frac{r_2}{4\beta_2 r_1^{2-\tau}} \min\left\{\|z\|^2, 1\right\}, \quad 0 < r_1(0) \leq 1 \\ \dot{r}_2 &= \frac{r_2^2}{2\alpha_2 r_1^{1-\tau}} \max\left\{\frac{\|z\|^{2\mu}}{r_2^{\mu+1}} - 1, 0\right\}, \qquad r_2(0) \geq \max\{8nc\|P\|, 1\}, \end{aligned} \tag{4.7}$$

where β_2 is given in (4.4), and $\mu \in (0, +\infty)$ is a constant to regulate the setting time.

Now we discuss the dynamics of r_1 and r_2. Because $r_2(0) \geq 1$ and $\dot{r}_2 \geq 0$, one has $r_2(t) \geq 1$. When $r_2 \geq \|z\|^{\frac{2\mu}{\mu+1}}$, we have $\dot{r}_2 = 0$ such that r_2 remains as a constant. Otherwise, since it holds that

$$\dot{r}_2 = \frac{r_2^2}{2\alpha_2 r_1^{1-\tau}} \left(\frac{\|z\|^{2\mu}}{r_2^{\mu+1}} - 1 \right),$$

r_2 is approaching $\|z\|^{\frac{2\mu}{\mu+1}}$. For r_1, when $\|z\| \geq 1$, it holds that

$$\dot{r}_1 = -\frac{r_2}{4\beta_2} \frac{1}{r_1^{2-\tau}} \left(r_1^2 - 1 \right).$$

Since $r_1(0) \in (0, 1]$, we can get $r_1(t) \in (0, 1)$. When $\|z\| \leq 1$, we have

$$\dot{r}_1 = -\frac{r_2}{4\beta_2 r_1^{2-\tau}} \left(r_1^2 - \|z\|^2 \right),$$

which means r_1 is approaching $\|z\|$. Therefore, it always holds that $r_1(t) \in [0, 1]$ and $r_2(t) \geq 1$.

4.4 STABILITY ANALYSIS

Now, we state the main result as follows:

Theorem 4.1

Suppose that Assumption 4.1 is satisfied. Let $K = (k_1, k_2, \ldots, k_n)$, P, and β_1, β_2 be determined through (4.3) and (4.4), then system (4.2) is globally fixed-time stabilized through the control law (4.5) when r_1, r_2 are designed as (4.7). Moreover, the setting time T can be chosen as

$$T = \frac{2 \left(\alpha_2 + \frac{\beta_1}{4} \right)^{\frac{1-\tau}{2}}}{\rho(1-\tau)} + \frac{2\alpha_2}{\alpha_1} + \frac{2\alpha_2^{\mu+1}}{\mu},$$

where $\rho = \min \left\{ \frac{1}{4(1+\tau)\alpha_2^{\frac{1+\tau}{2}}} \left(\frac{1+\tau}{\beta_2(1-\tau)} \right)^{\frac{1-\tau}{2}}, \frac{1}{2^{2-\tau}\beta_2^{\frac{1+\tau}{2}}} \right\}.$ ∎

Proof. Consider the state $z = (z_1, z_2, \ldots, z_n)^T$ in (4.6). Then, it satisfies

$$\dot{z} = \frac{r_2}{r_1^{1-\tau}} (A - BK)z - \frac{\dot{r}_1}{r_1} D_1 z - \frac{\dot{r}_2}{r_2} D_2 z + \tilde{F}, \tag{4.8}$$

where $D_1 = (n+1)I - (1-\tau)D$, $D_2 = D$, and

$$\tilde{F} = \begin{pmatrix} \frac{1}{r_1^{n+\tau} r_2} f_1(x_1) \\ \frac{1}{r_1^{n-1+2\tau} r_2^2} f_2(x_1, x_2) \\ \vdots \\ \frac{1}{r_1^{1+n\tau} r_2^n} f_n(x_1, x_2, \dots, x_n) \end{pmatrix}.$$

Letting $V = z^T P z$, we compute its derivative along with (4.8) as

$$\dot{V}|_{(4.8)} = \frac{r_2}{r_1^{1-\tau}} z^T \left(P(A - BK) + (A - BK)^T P \right) z - \frac{\dot{r}_1}{r_1} z^T (PD_1 + D_1 P) z$$
$$- \frac{\dot{r}_2}{r_2} z^T (PD_2 + D_2 P) z + 2z^T P\tilde{F}. \tag{4.9}$$

From (4.3) and (4.4), we have

$$0 \le PD_1 + D_1 P \le \beta_2 I,$$
$$0 \le PD_2 + D_2 P \le \beta_1 I,$$
$$\alpha_1 I \le P \le \alpha_2 I,$$

which implies

$$\dot{V}|_{(4.8)} \le + \frac{1}{4\beta_2} \frac{r_2}{r_1^{1-\tau}} z^T (PD_1 + D_1 P) z - \frac{r_2 \min\{\|z\|^2, 1\}}{4\beta_2 r_1^{3-\tau}} z^T (PD_1 + D_1 P) z$$
$$- \frac{r_2}{r_1^{1-\tau}} \|z\|^2 - \frac{r_2^2 \max\left\{ \frac{\|z\|^{2\mu}}{r_2^{\mu+1}} - 1, 0 \right\}}{2\alpha_2 r_1^{1-\tau}} z^T (PD_2 + D_2 P) z + 2z^T P\tilde{F} \tag{4.10}$$
$$\le - \frac{r_2}{r_1^{1-\tau}} \|z\|^2 + \frac{1}{4\beta_2} \frac{r_2}{r_1^{1-\tau}} z^T (PD_1 + D_1 P) z + 2z^T P\tilde{F}$$
$$\le - \frac{3r_2}{4r_1^{1-\tau}} \|z\|^2 + 2z^T P\tilde{F}.$$

To continue, we need to estimate the nonlinear term $\tilde{F}$. From Assumption 4.1, we know that

$$\left| \frac{1}{r_1^{n+1-i+i\tau} r_2^i} f_i(x_1, x_2, \dots, x_i) \right|$$
$$\le c \frac{|x_1|}{r_1^{n+1-i+i\tau} r_2^i} + c \frac{|x_2|}{r_1^{n+1-i+i\tau} r_2^i} + \dots + c \frac{|x_i|}{r_1^{n+1-i+i\tau} r_2^i}$$
$$\le c \frac{r_1^{(i-1)(1-\tau)}}{r_2^{i-1}} |z_1| + c \frac{r_1^{(i-2)(1-\tau)}}{r_2^{i-2}} |z_2| + \dots + c|z_i|$$
$$\le c\sqrt{n} \|z\|,$$

which indicates $\|\tilde{F}\| \leq cn\|z\|$. Back to (4.10), we can further get

$$\dot{V}|_{(4.8)} \leq -\frac{r_2}{2r_1^{1-\tau}}\|z\|^2 + 2cn\|P\|\|z\|^2$$

$$\leq -\frac{r_2}{2r_1^{1-\tau}}\|z\|^2 + \frac{2cn\|P\|}{r_1^{1-\tau}}\|z\|^2$$

$$\leq -\frac{r_2}{4r_1^{1-\tau}}\|z\|^2.$$

The following proof is divided into four parts. We first show the fixed-time attractivity of z. Then, we prove that z is fixed-time stable. After that, we derive the fixed-time stability of x from z. Finally, the upper bound of the setting time is estimated.

Part 1. Fixed-time attractivity of system (4.8):

In this part, we show that for any initial condition $z(0) \in \mathbb{R}^n$, $z(t)$ is dominated in the neighborhood $\Omega = \{z|\|z\| \leq 1\}$ before a setting time T_0.

Consider $\omega = \frac{V}{r_2}$. If $r_2 < \|z\|^{\frac{2\mu}{\mu+1}}$, r_2 in (4.7) satisfies

$$\dot{r}_2 = \frac{r_2^2}{2\alpha_2 r_1^{1-\tau}}\left(\frac{\|z\|^{2\mu}}{r_2^{\mu+1}} - 1\right),$$

and

$$\dot{\omega} = \frac{\dot{V}}{r_2} - \frac{\dot{r}_2}{r_2^2}V$$

$$\leq -\frac{1}{2r_1^{1-\tau}}\|z\|^2 - \frac{1}{2\alpha_2 r_1^{1-\tau}}\left(\frac{\|z\|^{2\mu}}{r_2^{\mu+1}} - 1\right)V \qquad (4.11)$$

$$\leq -\frac{1}{2\alpha_2 r_1^{1-\tau}}\frac{\|z\|^{2\mu}}{r_2^{\mu}}\frac{V}{r_2},$$

where $\|z\|^2 \geq \frac{V}{\alpha_2}$ is utilized.

If $r_2 \geq \|z\|^{\frac{2\mu}{\mu+1}}$, r_2 in (4.7) satisfies $\dot{r}_2 = 0$. The parameter r_2 will be a constant, and the derivative of ω is

$$\dot{\omega} \leq -\frac{1}{2\alpha_2 r_1^{1-\tau}}V \leq -\frac{1}{2\alpha_2 r_1^{1-\tau}}\frac{\|z\|^{2\mu}}{r_2^{\mu}}\frac{V}{r_2}. \qquad (4.12)$$

Thus, whether $r_2 < \|z\|^{\frac{2\mu}{\mu+1}}$ or $r_2 \geq \|z\|^{\frac{2\mu}{\mu+1}}$, it can be deduced from (4.11) and (4.12) that

$$\dot{\omega} \leq -\frac{1}{2\alpha_2 r_1^{1-\tau}}\frac{\|z\|^{2\mu}}{r_2^{\mu}}\frac{V}{r_2}.$$

Since $r_1 \leq 1$, we can further obtain $\dot{\omega} \leq -\frac{1}{2\alpha_2^{\mu+1}}\omega^{\mu+1}$, and

$$\omega(t) \leq \left(\frac{1}{\frac{1}{\omega^{\mu}(0)} + \frac{\mu}{2\alpha_2^{\mu+1}}t}\right)^{\frac{1}{\mu}}, \qquad \omega(0) > 0,$$

$$\omega(t) = 0, \qquad\qquad\qquad\qquad \omega(0) = 0.$$

When $t \geq \frac{2\alpha_2^{\mu+1}}{\mu}$, it holds that $\omega(t) \leq 1$. Back to (4.9), we obtain

$$\dot{V}|_{(4.8)} \leq -\frac{1}{2}\|z\|^2 V \leq -\frac{1}{2\alpha_2}V^2.$$

If $V(\frac{2\alpha_2^{\mu+1}}{\mu}) \neq 0$, one can get

$$V(t) \leq \frac{1}{\frac{1}{V(\frac{2\alpha_2^{\mu+1}}{\mu})} + \frac{1}{2\alpha_2}(t - \frac{2\alpha_2^{\mu+1}}{\mu})},$$

and

$$\|z(t)\|^2 \leq \frac{1}{\frac{\alpha_1}{V(\frac{2\alpha_2^{\mu+1}}{\mu})} + \frac{\alpha_1}{2\alpha_2}(t - \frac{2\alpha_2^{\mu+1}}{\mu})}.$$

Therefore, after the time instant $T_0 = \frac{2\alpha_2}{\alpha_1} + \frac{2\alpha_2^{\mu+1}}{\mu}$, $z(t)$ will be in the neighborhood Ω, i.e., $\|z(t)\|^2 \leq 1$ is satisfied. Meanwhile, by noticing $r_2 \geq 1$, after the time instant T_0, it holds that $\dot{r}_2(t) \equiv 0$, which ensures that $r_2(t)$ converges to a finite constant.

Part 2. Fixed-time stability of system (4.8):

When $t \geq T_0$, we have $\|z(t)\| \leq 1$. Under this condition, we find the instant T_1 such that $z(t) = 0$ for any $t \geq T_1$. Since $\frac{\|z\|^{2\mu}}{r_2^{\mu+1}} - 1 < 0$ holds, the parameter r_2 becomes constant. The parameter r_1 satisfies

$$\dot{r}_1 = -\frac{1}{4\beta_2}r_1^\tau r_2 + \frac{r_2}{4\beta_2 r_1^{2-\tau}}\|z\|^2.$$

Let $\varpi = V + \frac{\beta_2}{2}r_1^2$, and by utilizing (4.10), its derivative can be computed as

$$\dot{\varpi} \leq -\frac{r_2}{2r_1^{1-\tau}}\|z\|^2 - \frac{1}{4}r_1^{1+\tau}r_2 + \frac{r_2}{4r_1^{1-\tau}}\|z\|^2$$

$$\leq -\frac{1}{4r_1^{1-\tau}}\|z\|^2 - \frac{1}{4}r_1^{1+\tau}. \tag{4.13}$$

Employing Lemma 4.2 and choosing $\gamma = \left(\frac{1+\tau}{2(1-\tau)} \right)^{-\frac{1-\tau}{2}}$, we can get

$$-\frac{1}{4r_1^{1-\tau}} \|z\|^2 \le -\frac{\gamma^{-1}}{4(1+\tau)} \|z\|^{1+\tau} + \frac{(1-\tau)\gamma^{-\frac{2}{1-\tau}}}{4(1+\tau)} r_1^{1+\tau}.$$

From (4.13), it holds that

$$\dot{\varpi} \le -\frac{1}{4(1+\tau)\alpha_2^{\frac{1+\tau}{2}}} \gamma^{-1} V^{\frac{1+\tau}{2}} - \frac{1}{8} r_1^{1+\tau}$$

$$\le -\rho V^{\frac{1+\tau}{2}} - \rho \frac{\beta_2^{\frac{1+\tau}{2}}}{2^{1+\tau}} r_1^{1+\tau},$$

where $\rho = \min \left\{ \frac{1}{4(1+\tau)\alpha_2^{\frac{1+\tau}{2}}} \gamma^{-1}, \frac{1}{2^{2-\tau}\beta_2^{\frac{1+\tau}{2}}} \right\}$.

Thus, we can get

$$\dot{\varpi} \le -\rho(V + \frac{\beta_2}{4} r_1^2)^{\frac{1+\tau}{2}} \le -\rho \varpi^{\frac{1+\tau}{2}}, \tag{4.14}$$

where Lemma 4.1 is utilized. By considering inequality (4.14), we obtain

$$\varpi(t) \le \begin{cases} \left(\varpi^{\frac{1-\tau}{2}}(t_0) - \frac{1-\tau}{2} \rho(t-t_0) \right)^{\frac{2}{1-\tau}}, & T_0 \le t \le T, \\ 0, & t > T, \end{cases}$$

where $T = T_0 + \varpi^{\frac{1-\tau}{2}}(T_0) \frac{2}{(1-\tau)\rho}$. It can be further deduced that

$$\lim_{t \to T} \|z(t)\| = 0, \quad \lim_{t \to T} r_1(t) = 0.$$

Part 3. Fixed-time stability of system (4.2),(4.5),(4.7):
From the definition of z in (4.6), it holds that

$$x_1 = r_1^{n+\tau} r_2 z_1, \ x_2 = r_1^{n-1+2\tau} r_2^2 z_2, \ \ldots, \ x_n = r_1^{1+n\tau} r_2^n z_n.$$

Thus, from the convergence of z, we obtain the convergence of x as

$$\lim_{t \to T} \|x(t)\| = 0.$$

Meanwhile, from (4.10) we have

$$\|z(t)\|^2 \le \frac{1}{\alpha_1} V(t) \le \frac{1}{\alpha_1} V(0).$$

Because $r_1(t) \in [0,1]$ and $r_2(t) \leq \max\{r_2(0), \|z(t)\|^{\frac{2\mu}{\mu+1}}\}$, we have

$$\|x(t)\|^2 \leq \max\{r_2(0), \|z(t)\|^{\frac{2\mu}{\mu+1}}\}^{2n}\|z(t)\|^2$$

$$\leq \max\left\{r_2(0), \left(\frac{1}{\alpha}V(0)\right)\|^{\frac{\mu}{\mu+1}}\right\}^{2n}\left(\frac{1}{\alpha}V(0)\right).$$

It is noted that $V(0)$ is proportional to the initial state $\|x(0)\|$. Thus, the upper bound of $\|x(t)\|$ is governed by the initial state $\|x(0)\|$.

Part 4. Estimation of the setting time:

As discussed above, the FSC introduced in this paper contains three stages. The first stage is to ensure $\omega(t) \leq 1$, which happens when $t \geq \frac{2\alpha_2^{\mu+1}}{\mu}$. Then, under the condition $\omega(t) \leq 1$, we can get $\|z(t)\| \leq 1$ when $t \geq T_0$ with $T_0 = \frac{2\alpha_2}{\alpha_1} + \frac{2\alpha_2^{\mu+1}}{\mu}$. At last, from $\|z(t)\| \leq 1$, we have $\varpi(t) = 0$ when $t \geq T$ with $T = T_0 + \varpi^{\frac{1-\tau}{2}}(T_0)\frac{2}{(1-\tau)\rho}$. Since $\varpi(t_0) \leq \alpha_2 + \frac{\beta_1}{4}$, we have

$$T \leq \frac{2\left(\alpha_2 + \frac{\beta_1}{4}\right)^{\frac{1-\tau}{2}}}{\rho(1-\tau)} + \frac{2\alpha_2}{\alpha_1} + \frac{2\alpha_2^{\mu+1}}{\mu},$$

which is not dependent on the initial condition. Therefore, the control law (4.5) and (4.7) can render system (4.2) fixed-time stable for any initial state $x_0 \in \mathbb{R}^n$. This ends the proof. □

4.5　SIMULATION

In this section, we give two simulation examples to show the effectiveness of the control scheme developed in this chapter.

Example 4.1. *Consider the system*

$$\dot{x}_1 = x_2, \ \dot{x}_2 = x_3, \ \dot{x}_3 = u, \tag{4.15}$$

where $x = (x_1, x_2, x_3)^T$ is the system state, and u is the input.

From (4.5), the controller can be designed as

$$u = -\frac{r_2^{\frac{3}{2}}}{r_1^{\frac{3}{2}}}x_1 - 3\frac{r_2^{\frac{2}{2}}}{r_1}x_2 - 3\frac{r_2}{r_1^{\frac{1}{2}}}x_3, \tag{4.16}$$

where r_1, r_2 satisfy

$$\dot{r}_1 = -\frac{1}{4}r_1^{\frac{1}{2}}r_2 + \frac{r_2}{4r_1^{\frac{3}{2}}}\min\{\|z\|^2, 1\}, \quad r_1(0) = 1$$

$$\dot{r}_2 = \frac{1}{100}\frac{r_2^2}{2r_1^{\frac{1}{2}}}\max\left\{\frac{\|z\|^2}{r_2^2} - 1, 0\right\}, \quad r_2(0) = 1$$

with $z = \left(x_1/(r_1^{\frac{7}{2}}r_2), x_2/(r_1^3 r_2^2), x_3/(r_1^{\frac{5}{2}}r_2^3)\right)^T$.

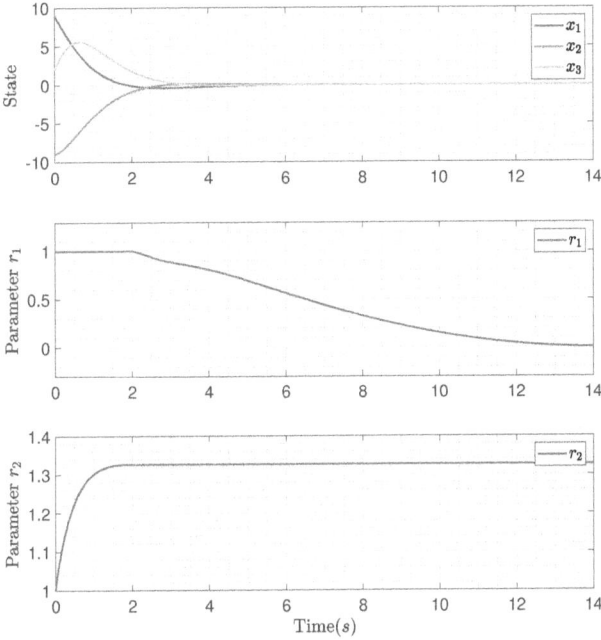

Figure 4.1 Trajectory for system (4.15), (4.16) under initial state $x(0) = (9, -9, 2)^T$.

We consider three different initial state conditions. Figures 4.1–4.3 show the simulation results under $x(0) = (9, -9, 2)^T$, $x(0) = (100, 10, -100)^T$ and $x(0) = (-0.2, 0.3, 0.1)^T$, respectively. From these figures, it is observed that r_1 converges to zero while r_2 increases to a constant for all these three cases. Moreover, all states in these three cases converge to zero before the setting time $T = 14\,s$. The simulation result is consistent with our theoretical analyses.

Example 4.2. *Consider the one-link manipulator system [69]. The system is described by*

$$D\ddot{q} + B\dot{q} + N\sin(q) = \tau_r,$$
$$M\dot{\tau}_r + H\tau_r = u - K_m\dot{q},$$
(4.17)

where q, $\dot{q}$, $\ddot{q}$ are the link position, velocity, and acceleration, respectively. τ_r is the torque produced by the electrical subsystem. u is the control input representing the electromechanical torque. D is the mechanical inertia, B is the coefficient of viscous friction at the joint, N is a positive constant related to the mass of the load and the coefficient of gravity, M is the armature inductance, H is the armature resistance, and K_m is the back electromotive force coefficient.

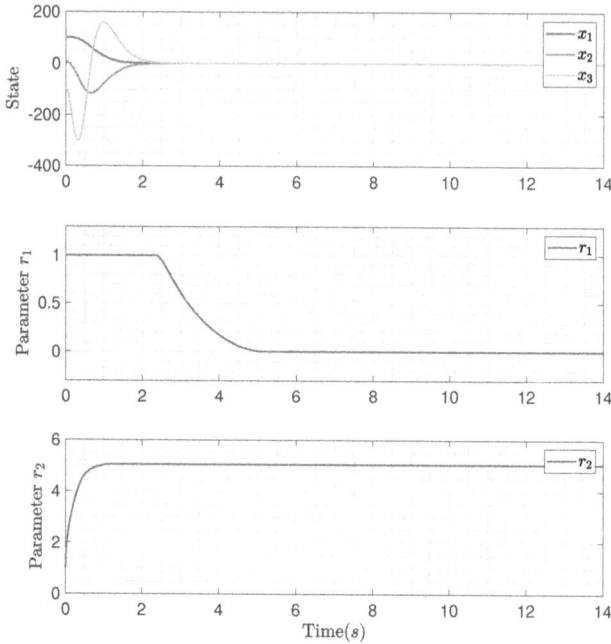

Figure 4.2 Trajectory for system (4.15), (4.16) under initial state $x(0) = (100, 10, -100)^T$.

Denoting $x_1 = MDq$, $x_2 = MD\dot{q}$, $x_3 = M\tau_r$, system (4.17) is transformed into

$$\begin{aligned}
\dot{x}_1 &= x_2, \\
\dot{x}_2 &= x_3 + f_2(x_1, x_2), \\
\dot{x}_3 &= u + f_3(x_2, x_3),
\end{aligned} \qquad (4.18)$$

where $f_2(x_1, x_2) = -\frac{B}{D}x_2 - MN\sin(\frac{x_1}{MD})$, $f_3(x_2, x_3) = -\frac{K_m}{MD}x_2 - \frac{H}{M}x_3$.

For simulation, the system parameters are set as $D = 5$ kg·m^2, $N = 0.2$, $M = 0.5$ H, $K_m = 0.05$ N·m/A, $B = 0.2$ N·ms/rad, and $H = 0.02$ Ω. It can be verified that Assumption 4.1 is satisfied with $c = 0.04$. Then, the control parameters are set as $k_1 = 2$, $k_2 = 6$, $k_3 = 3$, $\tau = 1/2$, $\beta = 20.7$, $\alpha_2 = 5$, $\mu = 3$. In particular, according to Theorem 4.1, the controller for this one-link manipulator system is designed as

$$u = -2r_2^3 r_1^{-\frac{3}{2}} x_1 - 6r_2^2 r_1^{-1} x_2 - 3r_2 r_1^{-\frac{1}{2}} x_3, \qquad (4.19)$$

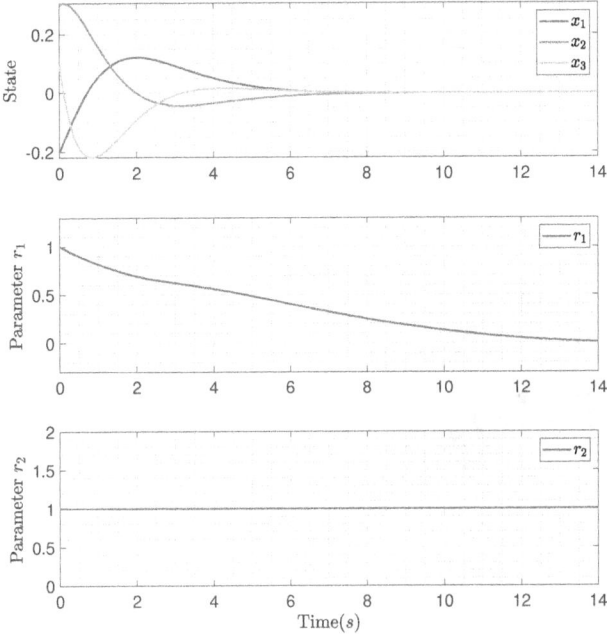

Figure 4.3 Trajectory for system (4.15), (4.16) under initial state $x(0) = (-0.2, 0.3, 0.1)^T$.

where r_1, r_2 are the dynamic gains designed as

$$\dot{r}_1 = -\frac{1}{4*20.7} r_1^{\frac{1}{2}} r_2 + \frac{r_2}{4*20.7 r_1^{\frac{3}{2}}} \min\left\{\|z\|^2, 1\right\},$$

$$\dot{r}_2 = \frac{r_2^2}{2*5 r_1^{\frac{1}{2}}} \max\left\{\frac{\|z\|^6}{r_2^4} - 1, 0\right\},$$

$$r_1(0) = 0.9, \quad r_2(0) = 2.62,$$

with $z = \left(x_1/(r_1^{\frac{7}{2}} r_2), \; x_2/(r_1^3 r_2^2), \; x_3/(r_1^{\frac{5}{2}} r_2^3)\right)^T$.

The simulation results are presented in Figure 4.4. It is observed that the system states x_1, x_2 and x_3 converge to zero before the time instant $T = 4$ s. Meanwhile, r_1 remains in the interval $[0, 1]$ and converges to zero, while r_2 increases to about 3.4 and then remains the same. The control signal u is also bounded and converges to zero. Moreover, it can be seen that the closed-loop system can still operate beyond the time interval $[0, 4]$, which cannot be achieved through the pre-specified finite-time control method in [117, 128].

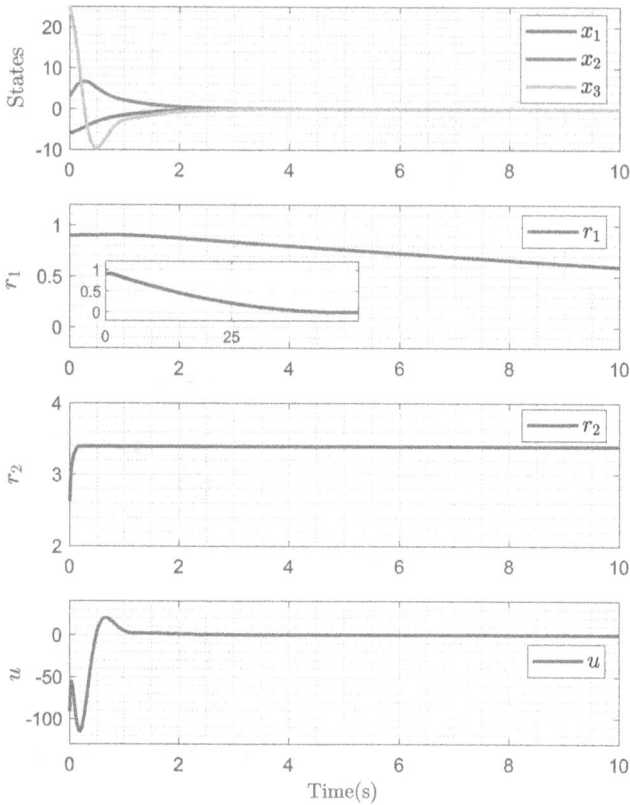

Figure 4.4 Simulation results for closed-loop system (4.18)-(4.20) with initial state $x(0) = (-6, 3, 25)^T$.

4.6 NOTES

In this chapter, the dynamic gain control approach has been used to solve the design problem of FSCs for nonlinear systems. An elegant solution to the considered problem is proposed by introducing two dynamic parameters. One dynamic parameter is used to regulate the system state to a pre-specified neighborhood, and the other dynamic parameter is designed to converge the system state to zero. Simulation studies have confirmed the effectiveness of the proposed method.

Part II

Variable Gain Control for
Feedforward Nonlinear Systems

5 Stabilization Control for Time-Varying Feedforward Nonlinear Systems

This chapter presents a variable gain control method for feedforward nonlinear systems. The idea of this method is to construct a controller with a low gain that can be updated online. The concerned system has an upper triangular structure, and the growth rate of the nonlinear dynamics is associated with a time-varying function. By introducing a suitable state transformation, the dynamic gain is properly designed which facilitates the design of the controller. By analyzing the convergence of the transformed systems, the stability of the original system is then determined. A simulation example is given to illustrate the effectiveness of the developed control method.

5.1 BACKGROUND

The control design of nonlinear feedforward systems has been extensively studied as they can model many practical systems, such as planar vertical takeoff and landing (PVTOL) aircraft and the inertia wheel pendulum. In general, two kinds of conditions for feedforward nonlinear systems are used to achieve stability analysis. The first is that the nonlinear system dynamics should be bounded by the sum of the quadratic or higher terms of its arguments. Under this condition, the stabilization problem through the saturation control method is studied in [122, 95, 136, 161]. The second one is that the nonlinear dynamics should satisfy the Lipschitz condition. This condition can involve a coefficient in different forms, such as a known constant [152], an unknown constant [9], a function dependent on the input or output [147, 58], and a function dependent on both the input and output [54].

For many practical systems, their models are time-varying, i.e., the models change over time. Mathematically, there are two main ways to characterize the slowly time-varying system models. First, their time derivatives are small enough. The unified stability criteria are studied in [28] for slowly time-varying systems under such a scenario. Second, the system models contain a slowly time-varying gain. If such a gain satisfies some growth requirements, the stability can be guaranteed for linear systems [50]. For feedforward nonlinear systems, the time-varying terms are always involved in the nonlinear functions, whose growth rate increases with time. Therefore, the existing results dealing with bounded nonlinear functions cannot be applied.

In this chapter, we study the problem of control design for a class of slowly time-varying feedforward nonlinear systems, in which the growth rate of the nonlinear functions is characterized by non-decreasing time-varying functions. By developing the low-gain feedback control technology, we introduce a novel dynamic gain to realize coordinate transformation and controller design. The stability of the closed-loop system is rigorously proved by Lyapunov analysis. Meanwhile, a simulation example is used to demonstrate the effectiveness of the proposed control strategy.

5.2 PROBLEM DESCRIPTION

The feedforward nonlinear system is given as

$$
\begin{cases}
\dot{x}_1 = x_2 + f_1\left(t, x_3, x_4, \ldots, x_n, u\right), \\
\dot{x}_2 = x_3 + f_2\left(t, x_4, \ldots, x_n, u\right), \\
\quad\vdots \\
\dot{x}_{n-1} = x_n + f_{n-1}\left(t, u\right), \\
\dot{x}_n = u,
\end{cases}
\tag{5.1}
$$

where $x = (x_1, x_2, \ldots, x_n)^T \in \mathbb{R}^n$ is the system state, $u \in \mathbb{R}$ is the control input, and $f_1, f_2, \ldots, f_{n-1}$ are continuous functions satisfying the following assumptions.

Assumption 5.1. *For any $x_1, x_2, \ldots, x_n, u \in \mathbb{R}$, it holds that*

$$
\left| f_i\left(t, x_{i+2}, \ldots, x_{n+1}\right) \right| \le \phi_1(t)\phi_2(u)\left(\sum_{j=i+2}^{n+1} |x_j| + |u| \right), \quad i = 1, \ldots, n-1,
\tag{5.2}
$$

where $x_{n+1} = 0$, $\phi_1(t)$ is a non-decreasing smooth function, and $\phi_2(u)$ is a continuous function.

Assumption 5.2. *For the continuous function $\phi_1(t)$, there exists a positive constant ε such that*

$$
\sup_{t \in [0,+\infty)} \dot{\phi}_1(t) \le \varepsilon.
\tag{5.3}
$$

Remark 5.1. *Assumption 5.2 implies that $\phi_1(t)$ is slowly time-varying as ε is small. For a linear system $\dot{x} = A(t)x$, this assumption means that $\|\dot{A}(t)\| \le \eta$ or $\int_t^{t+a} \|\dot{A}(s)\| ds \le \eta$ with a, η being positive constants [25, 27, 28]. When we consider the regulation problem, (5.3) can be replaced by $\limsup_{t \to +\infty} \dot{\phi}_1(t) \le \varepsilon$.*

Remark 5.2. *Inequality (5.2) is a generalized condition imposed on $f_i(\cdot)$. Some special cases of (5.2) include: 1) $\phi_1(t) = \theta$, $\phi_2(u) = 1$ with θ being a constant [152, 21, 9]; 2) $\phi_1(t) = \theta$, $\phi_2(u) = g(u)$ with θ being a constant and $g(u)$ being a function of u [147, 63]; and 3) $\phi(t) = (1 + t^\theta)$, $\phi_2(u) = 1$ with $\theta \in [0,1)$ [162].*

The problem to be addressed in this chapter is to design the input u such that (5.1) is globally asymptotically stable at the equilibrium $x = 0$.

5.3 CONTROL DESIGN

The controller u is given as

$$u = -\frac{k_1}{h^n}x_1 - \frac{k_2}{h^{n-1}}x_2 + \ldots - \frac{k_n}{h}x_n, \tag{5.4}$$

where the constants $k_1, k_2, \ldots, k_n$ and the dynamic gain h are to be designed later.

With the help of the state transformation

$$z_1 = \frac{x_1}{h^n}, z_2 = \frac{x_2}{h^{n-1}}, \ldots, z_n = \frac{x_n}{h},$$

we get

$$\begin{cases} \dot{z}_1 = \dfrac{z_2}{h} + \dfrac{f_1(x_3, x_4, \ldots, x_n, u)}{h^n} - n\dfrac{\dot{h}}{h}z_1, \\[2mm] \dot{z}_2 = \dfrac{z_3}{h} + \dfrac{f_2(x_4, \ldots, x_n, u)}{h^{n-1}} - (n-1)\dfrac{\dot{h}}{h}z_2, \\[2mm] \quad\vdots \\[2mm] \dot{z}_{n-1} = \dfrac{z_n}{h} + \dfrac{f_{n-1}(u)}{h^2} - 2\dfrac{\dot{h}}{h}z_{n-1}, \\[2mm] \dot{z}_n = \dfrac{u}{h} - \dfrac{\dot{h}}{h}z_n. \end{cases}$$

Meanwhile, the decreasing gain feedback input u in (5.4) can be converted into

$$u = -k_1 z_1 - k_2 z_2 - \ldots - k_n z_n.$$

Let $z = (z_1, z_2, \ldots, z_n)^T$, then u can be rewritten as $u = -Kz$ with $K = (k_1, k_2, \ldots, k_n)$. The closed-loop system can be described as

$$\dot{z} = \frac{1}{h}\tilde{A}z - \frac{\dot{h}}{h}Dz + F, \tag{5.5}$$

where

$$\tilde{A} = \begin{pmatrix} 0 & 1 & 0 & \cdots & 0 \\ 0 & 0 & 1 & \cdots & 0 \\ \vdots & \vdots & \vdots & & \vdots \\ 0 & 0 & 0 & \cdots & 1 \\ -k_1 & -k_2 & -k_3 & \cdots & -k_n \end{pmatrix}, \quad F = \begin{pmatrix} \frac{1}{h^n}f_1(x_3, x_4, \ldots, x_n, u) \\ \frac{1}{h^{n-1}}f_2(x_4, \ldots, x_n, u) \\ \vdots \\ \frac{1}{h^2}f_{n-1}(u) \\ 0 \end{pmatrix},$$

and $D = \text{diag}\{n, n-1, \ldots, 1\}$.

If K is properly chosen, we can find a positive definite matrix $P \in \mathbb{R}^{n \times n}$ satisfying

$$\tilde{A}^T P + P\tilde{A} \leq -I, \quad DP + PD \geq \alpha I. \tag{5.6}$$

Then, the dynamic gain h is designed as

$$h = (\phi_1(t) + 1) r, \tag{5.7}$$

where r satisfies

$$\dot{r} = \frac{1}{\alpha h} \max\left\{\beta \phi_2(u) - \frac{r}{2}, 0\right\}, \quad r(t_0) \geq 1, \tag{5.8}$$

with $\beta = 2\sqrt{n}\|P\|(\sqrt{n} + \|K\|)$. The function $\phi_2(u)$ is given in Assumption 5.1.

5.4 STABILITY ANALYSIS

The main result can be summarized as follows.

Theorem 5.1

Suppose that Assumptions 5.1–5.2 are satisfied. If ε satisfies (5.4), then the closed-loop system (5.1) under the control (5.4) is globally asymptotically stable. ∎

Proof. From (5.8), we have that h and r satisfy $h \geq 1$, $r \geq 1$, and $\dot{r} \geq 0$. Since $\dot{\phi}_1 \geq 0$, we have

$$\dot{h} \geq \frac{1}{\alpha h}\left(\beta(\phi_1(t) + 1)\phi_2(u) - \frac{h}{2}\right). \tag{5.9}$$

Consider the Lyapunov function candidate $V = z^T P z$. Then, we have

$$\dot{V}|_{(5.5)} \leq -\frac{1}{h}\|z\|^2 - \alpha\frac{\dot{h}}{h}\|z\|^2 + 2z^T PF, \tag{5.10}$$

where $\dot{h} \geq 0$, $h > 0$, and (5.6) is used.

From Assumption 5.1, it holds that

$$\frac{1}{h^{n+1-i}}|f_i| \leq \frac{1}{h^{n+1-i}}\phi_1(t)\phi_2(u)\left(|x_{i+2}| + \ldots + |x_n| + |u|\right)$$

$$\leq \frac{1}{h^2}\phi_1(t)\phi_2(u)\left(|z_{i+2}| + \ldots + |z_n| + |Kz|\right)$$

$$\leq \frac{1}{h^2}\phi_1(t)\phi_2(u)\left(\sqrt{n} + \|K\|\right)\|z\|,$$

for $i = 1, 2, \ldots, n-2$. It also holds that

$$\frac{1}{h^2}|f_{n-1}| \leq \phi_1(t)\phi_2(u)\frac{1}{h^2}\|K\|\,\|z\|.$$

Then, we obtain that

$$\|F\| \leq \phi_1(t)\phi_2(u)\frac{1}{h^2}\left(\sqrt{n} + \|K\|\right)\sqrt{n}\,\|z\|.$$

Substituting the above function into (5.10) gives

$$\dot{V}|_{(5.5)} \leq -\frac{1}{h}\|z\|^2 - \alpha\frac{\dot{h}}{h}\|z\|^2 + \beta\phi_1(t)\phi_2(u)\frac{1}{h^2}\|z\|^2.$$

From (5.9), we have

$$\dot{V}|_{(5.5)} \leq -\frac{1}{2h}\|z\|^2 \leq -\frac{1}{2h\lambda_{\max}(P)}V, \qquad (5.11)$$

where $\lambda_{\max}(P)$ is the maximum eigenvalue of P. Therefore, $z(t)$ in system (5.5) is bounded. Since $u = -Kz$, we have

$$|u|^2 \leq \|K\|^2\|z\|^2 \leq \frac{\|K\|^2}{\lambda_{\min}(P)}V(0)$$

with $\lambda_{\min}(P)$ being the minimum eigenvalue of P. Since $\phi_2(u)$ is a smooth function, we can find a constant $\bar{\phi}$ such that $\phi_2(u) \leq \bar{\phi}$. Thus, the time-varying constant $r(t)$ is bounded by a constant $\bar{r}$ and

$$\phi_1(t) + 1 \leq h \leq (\phi_1(t) + 1)\bar{r}.$$

Now, we analyze the convergence property of $x(t)$. Based on the definition of V, we have

$$\frac{\lambda_{\min}(P)}{(\phi_1(t)+1)^{2n}\bar{r}^{2n}}\|x\|^2 \leq \frac{\lambda_{\min}(P)}{h^{2n}}\|x\|^2 \leq V.$$

Thus, the convergence of x is determined by the term $\omega = (\phi_1(t)+1)^{2n+1}V$. From (5.11), we can obtain

$$\dot{\omega} \leq -\frac{1}{h\lambda_{\max}(P)}\omega + (2n+1)\frac{\bar{r}\dot{\phi}_1}{h(\phi_1+1)}\omega.$$

By choosing

$$\varepsilon \leq \frac{1}{2\lambda_{\max}(P)(2n+1)\bar{r}},$$

we obtain

$$\dot{\omega} \leq -\frac{1}{2h\lambda_{\max}(P)}\omega.$$

If $h(t)$ is bounded, $\lim_{t\to+\infty}\omega(t) = 0$. If $\lim_{t\to+\infty}h = +\infty$, we have that $\omega(t)$ is bounded. In both cases, when t tends $+\infty$, $\frac{1}{h}\omega(t) \to 0$ holds. With the relation

$$\|x\|^2 \leq \frac{h^{2n}}{\lambda_{\min}(P)}V \leq \frac{\omega}{h},$$

we can get

$$\lim_{t\to+\infty}\|x(t)\|^2 = 0,$$

which ends the proof. $\square$

In Theorem 5.1, the time-varying rate ε is not easily determined, since it is related to the upper bound of $\phi_2(u)$, i.e., $\bar{r}$. If we impose some extra conditions on $\phi_2(u)$, this problem can be relaxed. When $\phi_2(u)$ is bounded by a constant $\bar{\phi}_2$, we have the following result.

Theorem 5.2

Suppose that Assumption 5.1 is satisfied. If $\limsup_{r \to +\infty} \dot{\phi}_1(t) = 0$, then for any initial condition, the state of the system (5.1) is regulated to the equilibrium $x = 0$ by the control (5.4) with the dynamic parameter (5.7). ∎

Proof. From the proof of Theorem 5.1, by defining $\omega = (\phi_1(t) + 1)^{2n+1} z^T P z$, we can obtain

$$\dot{\omega} \leq -\frac{1}{h\lambda_{\max}(P)} \omega + (2n+1) \frac{\bar{r}\dot{\phi}_1}{h(\phi_1 + 1)} \omega.$$

Since $\limsup_{r \to +\infty} \dot{\phi}_1(t) = 0$, we can find an instant T_1 such that

$$\dot{\omega} \leq -\frac{1}{2h\lambda_{\max}(P)} \omega, \quad t \geq T_1.$$

Thus, a similar analysis can also deduce that

$$\lim_{t \to +\infty} \|x(t)\|^2 = 0,$$

which ends the proof. □

On the other hand, when the input u belongs to a closed set that depends on the initial states, a semi-global stability result can be obtained, as stated follow:

Theorem 5.3

Suppose all Assumptions of Theorem 5.1 are satisfied. Let X_0 be a closed set satisfying $X_0 \subset \{x \mid \|x\| \leq \rho\}$ for a positive constant ρ. Then, for any initial state $x(0) \in X_0$, the closed-loop system is stable when ε satisfies

$$\varepsilon \leq \frac{1}{8\lambda_{\max}(P)(2n+1)\beta\bar{\phi}_u},$$

where β is given in (5.8), and $\bar{\phi}_u = \sup_{u \in U_0} \phi_2(u)$ with $U_0 = \{u \mid \|u\| \leq \rho \|K\| \sqrt{\frac{\lambda_{\max}(P)}{\lambda_{\min}(P)}}\}$. ∎

Proof. As discussed before, we have $|u|^2 \leq \frac{\|K\|^2}{\lambda_{\min}(P)} V(0)$. Since

$$V(0) \leq \frac{\lambda_{\max}(P)}{r^2} \|x(0)\|^2 \leq \rho\lambda_{\max}(P),$$

we have

$$|u| \leq \rho\|K\| \sqrt{\frac{\lambda_{\max}(P)}{\lambda_{\min}(P)}}.$$

Thus, the semi-global stabilization result can be established due to the continuity of $\phi_2(u)$, which ends the proof. □

From the above discussion, it can be seen that the time-varying rate ε has to be small enough to stabilize a complex nonlinear system. When the system is time-invariant, we present the results without proof as follows.

Theorem 5.4

Suppose that Assumption 5.1 is satisfied with $\phi_1(t) = 1$. Then system (5.1) can be stabilized through the control

$$u = -\frac{k_1}{h^n}x_1 - \frac{k_2}{h^{n-1}}x_2 - \dots - \frac{k_n}{h}x_n,$$

where the dynamic gain h is designed as

$$h = \frac{1}{\alpha h} \max\left\{2\phi(u)(n + \sqrt{n}\|K\|)\|P\| - \frac{h}{2}, 0\right\}, \quad h(0) \geq 1$$

with $k_1, k_2, \dots, k_n$, and P given in (5.6). ∎

Theorem 5.5

Suppose that Assumption 5.1 is satisfied with $\phi_1(t) = 1$. Let X_0 be a closed set satisfying $\|X_0\| \subset \{x | \|x\| \leq \rho\}$ for a positive constant ρ. Then, for any initial state $x(0) \in X_0$, system (5.1) is stable through the control

$$u = -\frac{k_1}{h^n}x_1 - \frac{k_2}{h^{n-1}}x_2 - \dots - \frac{k_n}{h}x_n,$$

where $K = (k_1, k_2, \dots, k_n)$, positive definite matrix P satisfies $\tilde{A}^T P + P\tilde{A} \leq -I$, and the parameter h is a positive constant satisfying

$$h \geq \max\left\{4(\sqrt{n} + \|K\|)\sqrt{n}\|P\|\bar{\phi}_u, 1\right\}$$

with $\bar{\phi}_u = \sup_{u \in U_0} \phi_2(u)$ and $U_0 = \{u | |u| \leq \rho\|K\| \sqrt{\frac{\lambda_{\max}}{\lambda_{\min}}}\}$. ∎

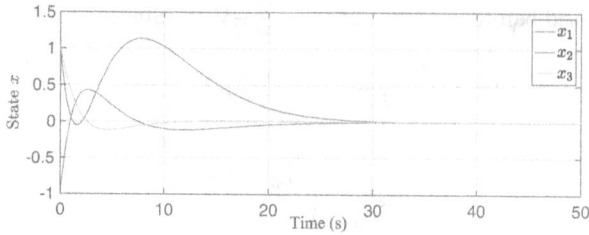

(a) $h = 3$ and $x(0) = (1, -1, 1)^T$.

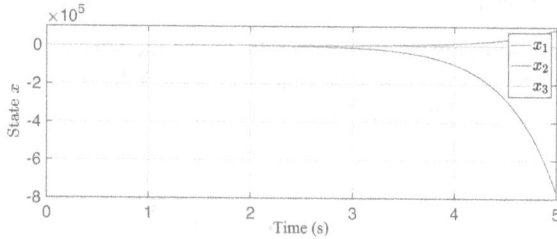

(b) $h = 3$ and $x(0) = (10, -10, 10)^T$.

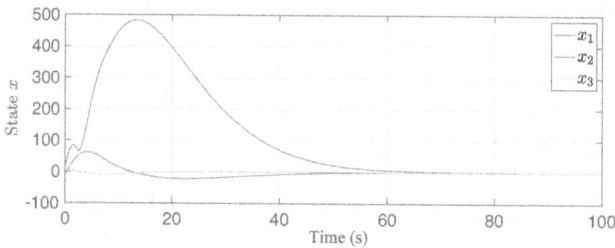

(c) $h = 6$ and $x(0) = (10, -10, 10)^T$.

Figure 5.1   System states under different h and $x(0)$.

5.5   SIMULATION

We consider a numerical time-varying system

$$\begin{cases} \dot{x}_1 = x_2 - \phi(t)u^2|x_3| - u(u^2 + 1), \\ \dot{x}_2 = x_3 + \phi(t)u^2, \\ \dot{x}_3 = u, \end{cases} \tag{5.12}$$

where $x = (x_1, x_2, x_3)^T \in \mathbb{R}^3$ is the system state, and $u \in \mathbb{R}$ is the control input.

It is easy to verify that system (5.12) satisfies Assumption 5.1 with $\phi_2(u) = \max\{u, u^2\} + 1$. We consider the following two cases to verify Theorem 5.2 and Theorem 5.5.

Part I: Verification of Theorem 5.5.

We assume that $\phi(t) = 1$. Then, following our result, the control law for system (5.12) can be designed as

$$u = -\frac{1}{h^3}x_1 - \frac{3}{h^2}x_2 - \frac{3}{h}x_3,$$

where h is a design parameter. We consider three different sets of parameters with the simulation results given in Figure 5.1. In Figure 5.1a, it is shown that the system states converge to zero with $h = 3$ and $x(0) = (1, -1, 1)^T$. When $x(0)$ increases to $(10, -10, 10)^T$, it is shown in Figure 5.1b that the gain $h = 3$ cannot stabilize the system states. However, if the gain increases to $h = 6$, the system stability is obtained again as shown in Figure 5.1c. The simulation result is consistent with the conclusion in Theorem 5.5.

Part II: Verification of Theorem 5.2.

We consider $\phi(t) = \ln(3t + 2)$. Then, it holds that

$$\lim_{t \to +\infty} \dot{\phi}(t) = \lim_{t \to +\infty} \frac{3}{3t + 2} = 0.$$

Thus, according to Theorem 5.2, the controller can be designed as

$$u = -\frac{1}{h^3}x_1 - \frac{3}{h^2}x_2 - \frac{3}{h}x_3,$$

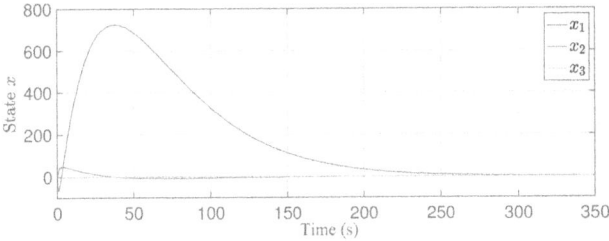

(a) The trajectory of x.

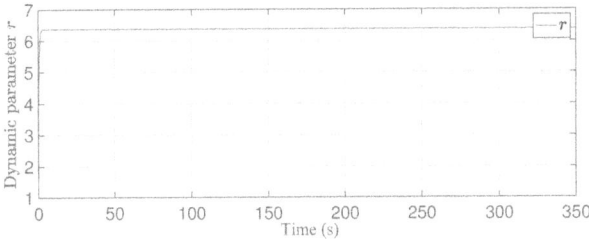

(b) The trajectory of r

Figure 5.2 The control performance under $x(0) = (10, -10, 10)^T$.

where h is a dynamic gain chosen as $h = (\ln(3t+2)+1)r$ with

$$\dot{r}(t) = \frac{1}{3h} \max \left\{ 2\max\{u, u^2\} + 2 - \frac{r(t)}{2}, 0 \right\} ds, \quad r(0) = 1.$$

The simulation result is shown in Figure 5.2. It is depicted in Figure 5.2a that the system states converge to zero. Figure 5.2b shows that the dynamic gain r converges to a constant, which verifies our result.

5.6   NOTES

In this chapter, we have studied the variable gain control problem for a class of nonlinear feedforward systems. In order to obtain a global result and dominate the time-varying terms, a novel dynamic gain is designed. A control strategy is then designed to ensure the stability of the system. In contrast to the existing works, the considered system is more general in terms of the description of the time-varying terms of the system. Simulation results have proven the control performance of the proposed control strategy.

6 Stabilization Control for Feedforward Nonlinear Time-Delay Systems

This chapter proposes the state feedback controller for a class of nonlinear systems with a delay in the input. The designed controller does not require saturation or recursive computation, which is widely used in the control design of feedforward nonlinear systems. With the help of coordinate transformation, the problem of controller design is reduced to the design of a gain parameter, which is essentially an optimization problem with linear matrix inequality (LMI) constraints. A simulation example is given to demonstrate the effectiveness of the proposed design procedure.

6.1 BACKGROUND

Time delays are frequently encountered in various technical systems. Therefore, the stability analysis of time-delay systems is of great theoretical and practical significance. Input delays often exist in practical systems due to the transmission of measurement signals. The existence of these delays can lead to instability or serious deterioration in the performance of the closed system [115]. Hence, the control of input delayed systems has attracted considerable attention. In [37], an H_∞ controller for a class of lower-triangular linear systems with input delay is studied based on the backstepping method. In [95], the problem of the global uniform asymptotic stabilization of a chain of integrators with input delays is solved.

Feedforward nonlinear systems are a class of widely studied systems. The problem of the asymptotic stabilization by state feedback of these triangular equations in the absence of delay has been studied by many researchers [105, 138]. The problems of global uniform asymptotic stability and local exponential stability were solved in [94] for a family of feedforward nonlinear systems when there is a delay in the input. The recursive design method has been shown to be effective in controlling feedforward systems without delay. However, the recursive design method cannot be directly extended to nonlinear feedforward systems with time delay, as the construction of a Krasovskii function or a Razimikhin function is non-trivial.

Motivated by [106, 102], we studied the design of controllers for a class of nonlinear feedforward systems with a delay in the input. Based on a suitable state transformation, the problem of controller design is converted into the design of a static gain. By constructing suitable Lyapunov-Krasovskii functions, a static gain design method is proposed in this chapter. The method used here is different from that used in [94], and is simpler because no recursive design procedure is used. Furthermore, the designed state feedback controller allows a large delay of the system input.

6.2 PROBLEM DESCRIPTION

Consider the following nonlinear system

$$\begin{cases} \dot{x}_1 = x_2 + \phi_1(t,x,u(t-d)), \\ \dot{x}_2 = x_3 + \phi_2(t,x,u(t-d)), \\ \quad \vdots \\ \dot{x}_{n-1} = x_n + \phi_{n-1}(t,x,u(t-d)), \\ \dot{x}_n = u(t-d), \end{cases} \tag{6.1}$$

where $x = (x_1, x_2, \ldots, x_n)^T \in \mathbb{R}^i$, represent the system states, u is the system input, and $d \geq 0$ denotes the delay. ϕ_i are continuous nonlinear functions satisfying the following Assumption 6.1.

The control objective here is to design a state feedback controller which can stabilize system (6.1) in the presence of time delays. To achieve the above control objective, the following assumption and lemma are given as follows:

Assumption 6.1. *There exist known positive constants c and c_1 such that*

$$|\phi_i(t,x,u(t-d))| \leq c(|x_{i+2}| + |x_{i+3}| + \ldots + |x_n|) + c_1|u(t-d)|, \ i = 1,2,\ldots,n-2,$$
$$|\phi_{n-1}(t,x,u(t-d))| \leq c_1|u(t-d)|.$$

For controller design, we also need the lemma given follows:

Lemma 6.1: [149]

For any continuous vector $\alpha \in R^n$ and symmetrical matrix $W > 0$, the following inequality holds

$$\int_{t-d}^{t} \alpha^T(\tau)d\tau W \int_{t-d}^{t} \alpha(\tau)d\tau \leq d \int_{t-d}^{t} \alpha^T(\tau)W\alpha(\tau)d\tau.$$

∎

6.3 STABILIZING FEEDBACK CONTROL

We present the main theorem of this chapter here:

Theorem 6.1

Under Assumption 6.1, for any given delay $d \geq 0$, there exists a linear state feedback controller $u(t-d) = K_d x(t-d)$ such that system (6.1) is globally asymptotically stable, where the row vector K_d is dependent on the delay $d \geq 0$. ∎

Proof. To prove this theorem, we need to consider the following four cases:

Case 1: When $n = 1$, $d = 0$, this result naturally holds.

Case 2: When $n = 1$, $d > 0$, we can choose an appropriate $L > 0$ such that

$$\dot{x}_1 = -\frac{1}{L} x_1(t - d) \tag{6.2}$$

is globally asymptotically stable at $x = 0$. Since

$$x_1(t - d) = x_1 - \int_{t-d}^{t} \dot{x}_1(\tau) d\tau.$$

We have

$$\dot{x}_1 = -\frac{1}{L}\left(x_1 - \int_{t-d}^{t} \dot{x}_1(\tau) d\tau\right) = -\frac{1}{L} x_1 - \frac{1}{L^2} \int_{t-d}^{t} x_1(\tau - d) d\tau. \tag{6.3}$$

If the solution of (6.3) with an initial condition in $[-2d, 0]$ is globally asymptotically stable, then the solution of (6.2) with an initial condition in $[-d, 0]$ is also globally asymptotically stable [38].

Choose a Lyapunov-Krasovskii function

$$V = \frac{1}{2} x_1^2 + \frac{1}{2L^2} \int_{-d}^{0} \int_{\theta+t-d}^{t} x_1^2(\tau) d\tau d\theta.$$

Then, we have

$$\begin{aligned}
\dot{V}|_{(6.3)} &= -\frac{1}{L} x_1^2 - \frac{1}{L^2} x_1 \int_{t-d}^{t} x_1(\tau - d) d\tau + \frac{d}{2L^2} x_1^2 - \frac{1}{2L^2} \int_{-d}^{0} x_1^2(\theta + t - d) d\theta \\
&\leq -\frac{1}{L} x_1^2 + \frac{d}{2L^2} x_1^2 + \frac{1}{2L^2} \int_{t-d}^{t} x_1^2(\tau - d) d\tau + \frac{d}{2L^2} x_1^2 - \frac{1}{2L^2} \int_{-d}^{0} x_1^2(\theta + t - d) d\theta \\
&= -\frac{1}{L}\left(1 - \frac{d}{L}\right) x_1^2,
\end{aligned}$$

where the Schwarz inequality is applied.

Choose

$$L \geq \frac{d}{1 - \eta},$$

where η satisfies $0 < \eta < 1$, then

$$\dot{V}|_{(6.3)} \leq -\frac{\eta}{L} x_1^2 < 0.$$

Therefore, the solution of (6.3) is globally asymptotically stable, and hence (6.2) is also globally asymptotically stable.

Case 3: When $n \geq 2$, $d > 0$, we introduce the following change of coordinates

$$z = Nx, \tag{6.4}$$

where $N = \mathrm{diag}[1, L, \ldots, L^{n-1}]$, $L \geq 1$ is a constant to be determined later. Then, system (6.1) is converted into

$$\dot{z} = \frac{1}{L} J_1 z + \Phi + H_1, \tag{6.5}$$

where

$$J_1 = \begin{pmatrix} 0 & 1 & 0 & \cdots & 0 \\ 0 & 0 & 1 & \cdots & 1 \\ \vdots & \vdots & \vdots & \ddots & \vdots \\ 0 & 0 & 0 & \cdots & 1 \\ 0 & 0 & 0 & \cdots & 0 \end{pmatrix}, \quad \Phi = \begin{pmatrix} \phi_1 \\ L\phi_2 \\ \vdots \\ L^{n-2}\phi_{n-1} \\ 0 \end{pmatrix}, \quad H_1 = \begin{pmatrix} 0 \\ 0 \\ \vdots \\ 0 \\ L^{n-1}u(t-d) \end{pmatrix}.$$

Let $b_i > 0$, $i = 1, 2, \ldots, n$, be the coefficients of the Hurwitz polynomial

$$q(s) = s^n + b_n s^{n-1} + \ldots + b_2 s + b_1.$$

Now, we need to find an $L \geq 1$ such that the closed-loop system (6.5) and

$$u(t-d) = -\frac{1}{L^n}\big(b_1 z_1(t-d) + b_2 z_2(t-d) + \ldots + b_n z_n(t-d)\big) \tag{6.6}$$

is globally asymptotically stable at $z = 0$. Since

$$z_i(t-d) = z_i - \int_{t-d}^t \dot{z}_i(\tau) d\tau,$$

from (6.6), it follows that

$$u(t-d) = -\frac{1}{L^n}(b_1 z_1 + b_2 z_2 + \ldots + b_n z_n)$$
$$+ \frac{1}{L^n}\int_{t-d}^t \big(b_1 \dot{z}_1(\tau) + b_2 \dot{z}_2(\tau) + \ldots + b_n \dot{z}_n(\tau)\big) d\tau.$$

Then, (6.5) and (6.6) can be re-written as

$$\dot{z} = \frac{1}{L} Az + \Phi + H_2, \tag{6.7}$$

where

$$A = \begin{pmatrix} 0 & 1 & \cdots & 0 \\ 0 & 0 & \cdots & 1 \\ \vdots & \vdots & \ddots & \vdots \\ 0 & 0 & \cdots & 1 \\ -b_1 & -b_2 & \cdots & -b_n \end{pmatrix}, \quad H_2 = \begin{pmatrix} 0 \\ 0 \\ \vdots \\ 0 \\ \frac{1}{L}\int_{t-d}^t \big(b_1\dot{z}_1(\tau) + \ldots + b_n\dot{z}_n(\tau)\big) d\tau \end{pmatrix}.$$

Since $q(s)$ is a Hurwitz polynomial, we have that A is a stable matrix and there exists a positive definite matrix $P > 0$ such that

$$PA + A^T P = -I_n. \tag{6.8}$$

Let p_{ij} denote the element of P in the i-th row and j-th column.
Noting Assumption 6.1, (6.4) and $L \geq 1$, we can get

$$|L^{i-1}\phi_i| \leq \frac{c}{L^2}(|z_{i+2}|+\ldots+|z_n|) + \frac{c_1}{L^{n-i+1}}(b_1|z_1(t-d)|+\ldots+b_n|z_n(t-d)|)$$

for $i = 1, 2, \ldots, n-2$, and

$$|L^{n-1}\phi_n| \leq \frac{c}{L^2}(b_1|z_1(t-d)|+\ldots+b_n|z_n(t-d)|).$$

Choose $V_1 = z^T P z$, then

$$\dot{V}_1|_{(6.7)} = -\frac{1}{L}\|z\|^2 + 2z^T P\Phi + 2z^T PH_2. \tag{6.9}$$

For $2z^T P\Phi$, we have

$$2z^T P\Phi \leq \frac{2c}{L^2}|z|^T|P||J_2||z| + \frac{c_1}{L^2}|z|^T|P||J_3 R_0^{-1}J_3^T||z| + \frac{c_1}{L^2}|z(t-d)|^T R_0|z(t-d)|,$$

where

$$J_2 = \begin{pmatrix} 0 & 0 & 1 & 1 & \cdots & 1 & 1 \\ 0 & 0 & 0 & 1 & \cdots & 1 & 1 \\ \vdots & \vdots & \vdots & \vdots & \ddots & \vdots & \vdots \\ 0 & 0 & 0 & 0 & \cdots & 1 & 1 \\ 0 & 0 & 0 & 0 & \cdots & 0 & 1 \\ 0 & 0 & 0 & 0 & \cdots & 0 & 0 \\ 0 & 0 & 0 & 0 & \cdots & 0 & 0 \end{pmatrix}$$

$$J_3 = (1, 1, \ldots, 1, 1, 0)^T (b_1, b_2, \ldots, b_n), \quad R_0 > 0.$$

Noting $L \geq 1$, we have

$$|b_1\dot{z}_1 + b_2\dot{z}_2 + \ldots + b_n\dot{z}_n|$$
$$\leq \frac{1}{L}(b_1, \ldots, b_n)(J_1 + cJ_2)(|z_1|, |z_2|, \ldots, |z_n|)^T + \frac{1}{L}(b_1, \ldots, b_n)$$
$$\times (c_1, \ldots, c_1, 1)^T \left((b_1, \ldots, b_n)(|z_1(t-d)|, |z_2(t-d)|, \ldots, |z_n(t-d)|)^T\right).$$

Hence, based on Lemma 6.1, we have

$$2z^T PH_2 \leq \frac{d}{L^2}|z|^T(G_1 R_1^{-1}G_1^T + G_2 R_2^{-1}G_2^T)|z| + \frac{1}{L^2}\int_{t-d}^t |z(\tau)|^T R_1|z(\tau)|d\tau$$
$$+ \frac{1}{L^2}\int_{t-d}^t |z(\tau-d)|^T R_2|z(\tau-d)|d\tau, \tag{6.10}$$

where $R_1 > 0$, $R_2 > 0$ and

$$G_1 = \begin{pmatrix} |p_{1n}| \\ \vdots \\ |p_{nn}| \end{pmatrix}(b_1, \ldots, b_n)(J_1 + cJ_2), \quad G_2 = \begin{pmatrix} |p_{1n}| \\ \vdots \\ |p_{nn}| \end{pmatrix}(b_1, \ldots, b_n)\begin{pmatrix} c_1 \\ \vdots \\ c_1 \\ 1 \end{pmatrix}(b_1, \ldots, b_n).$$

It then follows from (6.9)-(6.10) that

$$
\begin{aligned}
\dot{V}_1|_{(6.7)} = & -\frac{1}{L}\|z\|^2 + \frac{1}{L^2}|z|^T (2c|P|J_2 + dG_1R_1^{-1}G_1^T + dG_2R_2^{-1}G_2^T)|z| \\
& + \frac{c_1}{L^2}|z|^T|P||J_3R_0^{-1}J_3^T||P|^T|z| + \frac{c_1}{L^2}|z(t-d)|^T R_0|z(t-d)| \\
& + \frac{1}{L^2}\int_{-d}^{0}|z(\theta+t)|^T R_1|z(\theta+t)|d\theta \\
& + \frac{1}{L^2}\int_{-d}^{0}|z(\theta+t-d)|^T R_2|z(\theta+t-d)|d\theta
\end{aligned}
\tag{6.11}
$$

Choosing a Lyapunov-Krasovskii function

$$
\begin{aligned}
V = & z^T Pz + \frac{c_1}{L^2}\int_{t-d}^{t}|z(s)|^T R_0|z(s)|ds + \frac{1}{L^2}\int_{-d}^{0}\int_{\theta+t}^{t}|z(s)|^T R_1|z(s)|dsd\theta \\
& + \frac{1}{L^2}\int_{-d}^{0}\int_{\theta+t-d}^{t}|z(s)|^T R_2|z(s)|dsd\theta,
\end{aligned}
$$

and noting (6.11), one can get

$$
\dot{V}_1|_{(6.7)} \le -\frac{1}{L}\|z\|^2 + \frac{1}{L^2}|z|^T W|z| = -\frac{1}{L^2}|z|^T(LI_n - W)|z|,
$$

where

$$
\begin{aligned}
W = & c|P|J_2 + cJ_2^T|P| + dG_1R_1^{-1}G_1^T + dG_2R_2^{-1}G_2^T \\
& + dR_1 + dR_2 + c_1|P||J_3R_0^{-1}J_3^T||P|^T + c_1R_0.
\end{aligned}
$$

Choosing $L \ge 1$ such that $LI_n - W > 0$, we can get $\dot{V}_1|_{(6.7)} \le 0$, which indicates that system (6.7) is globally asymptotically stable. Therefore, systems (6.5) and (6.6) are also globally asymptotically stable. Since the constant matrix W is independent of L, we can always find an L satisfying $LI_n - W > 0$.

From Schur complements, we know that finding $L \ge 1$ (as small as possible) that satisfies $LI_n - W > 0$ is equivalent to solving the following optimization problem with linear matrix inequality (LMI) constraints

$$
\min_{R_0,R_1,R_2} L
\tag{6.12}
$$

$$
s.t.\ L > 1 \text{ and } \begin{pmatrix} \Xi & \Omega \\ \Omega^T & \Lambda \end{pmatrix} > 0,
\tag{6.13}
$$

where

$$
\begin{aligned}
& \Xi = LI_n - c_1R_0 - dR_1 - dR_2 - c|P|J_2 - cJ_2^T|P|, \\
& \Omega = (\sqrt{c_1}|P|J_3, \sqrt{d}G_1, \sqrt{d}G_2), \\
& \Lambda = \text{diag}[R_0, R_1, R_2].
\end{aligned}
$$

The optimization problem (6.12) with constraint (6.13) can be easily solved by the LMI toolbox in MATLAB. Different from [133], the linear matrix inequality (6.13) always has a solution.

Case 4: When $n \geq 2$, $d = 0$, we can set H_2 and W in case 3 as

$$H_2 = 0, \quad W = c|P|J_2 + cJ_2^T|P| + c_1|P||J_3 R_0^{-1} J_3^T|P|^T + c_1 R_0.$$

Then, following the proof in case 3, similar conclusions can be drawn. The proof ends here.   □

Remark 6.1. *For any given time delay $d > 0$, we can always find a constant L satisfying (6.13). The purpose of finding a small L satisfying (6.13) is that we would like to get a relatively big control gain. In other words, there may be many sets b_i, $i = 1, 2, \ldots, n$, satisfying our design requirements, but the desired set is the one that can make L as small as possible. As a result, a large control gain is achieved such that the convergence rate of the closed-loop system is fast.*

6.4 SIMULATION

Consider the following nonlinear system

$$\begin{cases} \dot{x}_1 = x_2 - 0.5(x_3)^{\frac{1}{3}}(x_4)^{\frac{2}{3}}, \\ \dot{x}_2 = x_3 - \frac{1}{2+2e^{-t}}x_4, \\ \dot{x}_3 = x_4, \\ \dot{x}_4 = u(t-d). \end{cases} \tag{6.14}$$

Obviously, Assumption 6.1 holds for this system with $c = 0.5$ and $c_1 = 0$. For the sake of simulation, the time delay is set as $d = 0.1$. Choose $b_1 = 3.8$, $b_2 = 7.1$, $b_3 = 11.4$, $b_4 = 2.6$, then

$$A = \begin{pmatrix} 0 & 1 & 0 & 0 \\ 0 & 0 & 1 & 0 \\ 0 & 0 & 0 & 1 \\ -3.8 & -7.1 & -11.4 & -2.6 \end{pmatrix}.$$

Solving the Lyapunov equation (6.8), we can get

$$P = \begin{pmatrix} 2.6938 & 2.7877 & 1.5681 & 0.1316 \\ 2.7877 & 6.1167 & 3.3631 & 0.4631 \\ 1.5681 & 3.3631 & 4.0961 & 0.3389 \\ 0.1316 & 0.4631 & 0.3389 & 0.3226 \end{pmatrix} > 0.$$

By solving the optimization problems (6.12)–(6.13), we have $L = 13.6678$. So the state feedback controller for (6.14) is

$$u(t-d) = -\frac{1}{L^4} 3.8 x_1(t-d) - \frac{1}{L^3} 7.1 x_2(t-d)$$

$$- \frac{1}{L^2} 11.4 x_3(t-d) - \frac{1}{L} 2.6 x_4(t-d). \tag{6.15}$$

Figures 6.1–6.2 show the state response of the closed-loop systems (6.14) and (6.15) with $d = 0.1$ and the initial condition $(x_1(t), x_2(t), x_3(t), x_4(t)) = (15, 12, -1, -0.5)$, for $t \in [-0.1, 0]$. From these two figures, it is shown that system states are bounded and converge to the origin.

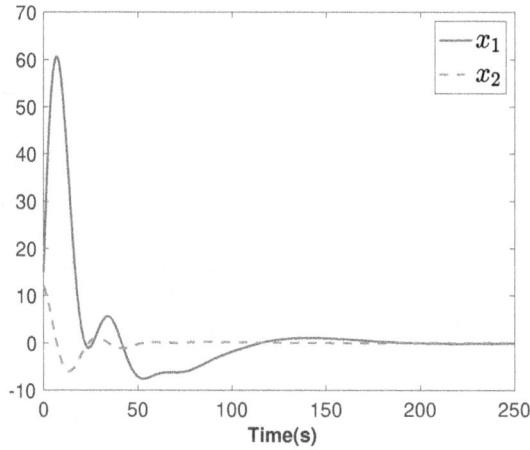

Figure 6.1   Trajectories of x_1 and x_2.

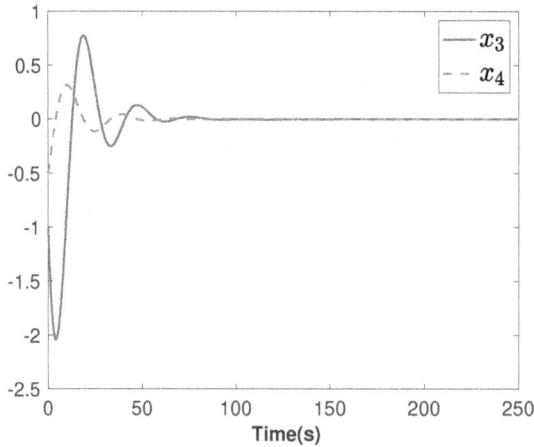

Figure 6.2   Trajectories of x_3 and x_4.

6.5   NOTES

In this chapter, we have studied the problem of global stabilization of a class of nonlinear feedforward systems with a delay in the input. Through coordinate transformation, the controller design is converted into an LMI optimization problem. The designed state feedback controller has a very simple structure without saturations or recursive calculations and is therefore easy to implement in practice.

7 Variable Gain Control for Discrete-Time Feedforward Nonlinear Systems

This chapter considers the stabilization control problem for a class of discrete-time feedforward nonlinear systems. The nonlinear dynamics in the systems considered are bounded by an input-dependent function. We first present a control method with a static gain to achieve semi-global stabilization. The criteria for setting the gain value are explicitly given based on Lyapunov stability analysis. Then, we propose a modified control method with a dynamic gain to achieve global stabilization. Finally, simulation results are provided to illustrate the effectiveness of the proposed methods.

7.1 BACKGROUND

Feedforward nonlinear systems (FNSs) are an important kind of nonlinear system that can model many practical systems. In particular, discrete-time FNSs are commonly used to model the planar vertical take-off and landing aircraft, inertial wheel pendulums, and active magnetic bearings with low-bias [135]. Although many results on the stabilization of continuous-time FNSs are available, the corresponding results on discrete-time FNSs are still limited [1, 92, 135].

There are some interesting results of studying the stabilization control of DFNSs. For example, a control strategy is proposed in [1] and [92] which can ensure the global stability of a class of DFNSs. The saturation methodology is used to construct the control law in [135], where the system nonlinearities satisfy a quadratic growth condition. In addition to the above methods, the low-gain control method is also widely applied in the control of continuous-time FNSs. It introduces a parameter into the control gain, and system stability can be ensured by adjusting the value of this parameter [76, 119, 129, 158]. However, the results of developing a low-gain feedback control for DFNSs are still very limited, which inspires the work of this chapter.

In this chapter, we study the control problem for a class of DFNSs, where the nonlinear functions have an input-dependent growth rate. First, we introduce a low-gain control method to realize semi-global stabilization. Then, a dynamic gain control method is proposed to achieve global stabilization. The stability of the closed-loop system is rigorously proved by Lyapunov analysis. Finally, simulation examples are given to illustrate the effectiveness of the proposed methods.

7.2   PROBLEM DESCRIPTION

The DFNS to be controlled in this chapter is given as

$$
\begin{aligned}
x_1(k+1) &= x_1(k) + p_1 x_2(k) + f_1(x(k), u(k)),\\
x_2(k+1) &= x_2(k) + p_2 x_3(k) + f_2(x(k), u(k)),\\
&\vdots\\
x_n(k+1) &= x_n(k) + p_n u(k) + f_n(x(k), u(k)),
\end{aligned}
\tag{7.1}
$$

where $x = (x_1, x_2, \ldots, x_n)^T \in \mathbb{R}^n$ is the system state, $u \in \mathbb{R}$ is the system input, and $p_1, p_2, \ldots, p_n$ are positive constants. The initial time instant is set as 0, and the initial state $x(0)$ belongs to the set $\Omega \in \mathbb{R}^n$. $f_1(\cdot), f_2(\cdot), \ldots, f_n(\cdot)$ are continuous functions satisfying the following assumption.

Assumption 7.1. *For $x = (x_1, x_2, \ldots, x_n) \in \mathbb{R}^n$, $u \in \mathbb{R}$, it holds that*

$$
\begin{aligned}
|f_i(x,u)| &\le \phi(u)(|x_{i+2}| + |x_{i+3}| + \ldots + |u|), \quad i = 1, 2, \ldots, n-2,\\
|f_{n-1}(x,u)| &\le \phi(u)|u|, \quad \text{and} \quad f_n(x,u) = 0
\end{aligned}
$$

where $\phi(u)$ is a continuous function with respect to u.

The control objective is to design the control input u such that (7.1) is asymptotically stable at the equilibrium $x = 0$, i.e., the system state $x(k)$ satisfies

$$
\lim_{k \to +\infty} \|x(k)\| = 0, \quad x(0) \in \Omega.
$$

Let

$$
A = \begin{pmatrix}
0 & p_1 & 0 & \cdots & 0\\
0 & 0 & p_2 & \cdots & 0\\
\vdots & \vdots & \vdots & & 0\\
0 & 0 & 0 & \cdots & p_{n-1}\\
0 & 0 & 0 & \cdots & 0
\end{pmatrix}, \quad
B = \begin{pmatrix}
0\\
0\\
\vdots\\
0\\
p_n
\end{pmatrix},
\tag{7.2}
$$

then system (7.1) can be rewritten in the matrix form as

$$
x(k+1) = (I+A)x(k) + Bu(k) + F(x(k), u(k)),
$$

where $F(\cdot) = (f_1(\cdot), f_2(\cdot), \ldots, f_n(\cdot))^T$. For matrices A and B in (7.2), there exist a vector K and a positive definite matrix P satisfying that

$$
(A - BK)^T P + P(A - BK) \le -\alpha_1 P,
\tag{7.3}
$$

and

$$
PD + DP \ge \alpha_2 P,
\tag{7.4}
$$

where $D = \mathrm{diag}\{n, n-1, \ldots, 1\}$, α_1 and α_2 are positive constants.

Lemma 7.1

For any positive definite matrix P satisfying (7.3)-(7.4), we have

$$\Gamma^T P \Gamma \le \frac{1}{\gamma^{\alpha_2}} P, \tag{7.5}$$

where $\Gamma = \text{diag}\left\{\frac{1}{\gamma^n}, \frac{1}{\gamma^{n-1}}, \frac{1}{\gamma^{n-2}}, \ldots, \frac{1}{\gamma}\right\}$ with $\gamma \ge 1$. ∎

Proof. Denote

$$\omega(v, \gamma) = v^T \Gamma P \Gamma v,$$

where $v \in \mathscr{V} = \{v \in \mathbb{R}^n | \|v\| = 1\}$.

Since P is a positive definite matrix, $\omega \ge 0$ holds for any $v \in \mathscr{V}$ and $\gamma \ge 1$. Meanwhile, we have

$$\frac{\partial \omega}{\partial \gamma} = -\frac{1}{\gamma} v^T \Gamma (DP + PD) \Gamma v \le -\frac{\alpha_2}{\gamma} \omega \le 0.$$

Using the Grönwall's inequality gives

$$\omega(v, \gamma) \le \frac{1}{\gamma^{\alpha_2}} \omega(v, 1)$$

for any $\gamma \ge 1$. By recalling $\omega(v, 1) = v^T P v$, we can obtain (7.5) which ends the proof. □

7.3 DESIGN OF LOW-GAIN FEEDBACK CONTROL

In this section, we introduce a static gain feedback control technology for DFNSs. First, consider the state transformation

$$z_1(k) = \frac{x_1(k)}{h^n}, z_2(k) = \frac{x_2(k)}{h^{n-1}}, \ldots, z_n(k) = \frac{x_n(k)}{h},$$

where the static gain h is to be designed later. Then, we have

$$z_1(k+1) = z_1(k) + p_1 \frac{1}{h} z_2(k) + \frac{1}{h^n} f_1(x(k), u(k)),$$

$$z_2(k+1) = z_2(k) + p_2 \frac{1}{h} z_3(k) + \frac{1}{h^{n-1}} f_2(x(k), u(k)),$$

$$\vdots$$

$$z_n(k+1) = z_n(k) + p_n \frac{1}{h} u(k) + \frac{1}{h} f_n(x(k), u(k)).$$

Construct the low-gain feedback controller as

$$u(k) = -Kz(k), \tag{7.6}$$

where K is the vector given in (7.3) and $z = (z_1, z_2, \ldots, z_n)^T$.

Subsequently, the closed-loop system can be written as

$$z(k+1) = \left(I + \frac{1}{h}A - \frac{1}{h}BK\right)z(k) + \tilde{F}(k), \tag{7.7}$$

where $\tilde{F}(k) = \left(\frac{1}{h^n}f_1(x(k),u(k)),\ldots,\frac{1}{h}f_n(x(k),u(k))\right)^T$.

Since h is a constant, the stabilization of (7.7) is equivalent to that of (7.1). Therefore, the control design is reduced to the design of h. The main result of this section is stated as follows.

Theorem 7.1

Suppose that Assumption 7.1 is satisfied. Then, system (7.1) with the controller (7.6) is stable if $x(0) \in \Omega$ with Ω being a closed set in $\mathbb{R}^n$ and $h \geq h_\Omega$ with h_Ω specified in (7.11). ∎

Proof. Consider the Lyapunov function candidate $V = z^T P z$. Then, we have

$$
\begin{aligned}
&V(k+1) - V(k) \\
={}& z^T(k+1)Pz(k+1) - z^T(k)Pz(k) \\
={}& \frac{1}{h}z^T(k)\left((A-BK)^T P + P(A-BK)\right)z(k) \\
&+ \frac{1}{h^2}z^T(k)(A-BK)^T P(A-BK)z(k) \\
&+ 2z^T(k)\left(I + \frac{1}{h}A - \frac{1}{h}BK\right)^T P\tilde{F}(k) \\
\leq{}& -\frac{\alpha_1}{h}z^T(k)Pz(k) + \frac{\beta_1}{h^2}z^T(k)Pz(k) \\
&+ 2\|P\|\left(\|I\| + \|A - BK\|\right)\|z(k)\|\|\tilde{F}(k)\|,
\end{aligned} \tag{7.8}
$$

where β_1 is a positive constant satisfying $(A-BK)^T P(A-BK) \leq \beta_1 P$.

Suppose $h \geq 1$, then from Assumption 7.1 we have

$$
\begin{aligned}
&\left|\frac{1}{h^{n+1-i}}f_i(x(k),u(k))\right| \\
\leq{}& \frac{\phi(u(k))}{h^{n+1-i}}\left(h^{n-1-i}|z_{i+2}(k)| + \ldots + h|z_n(k)| + |Kz(k)|\right) \\
\leq{}& \frac{1}{h^2}\phi(u(k))\left(\sqrt{n} + \|K\|\right)\|z(k)\|,
\end{aligned} \tag{7.9}
$$

for $i = 1,2,\ldots,n-1$, and $f_n(x(k),u(k)) = 0$. Thus, we can get

$$\|\tilde{F}(k)\| \leq \frac{1}{h^2}\phi(u(k))\left(n + \|K\|\sqrt{n}\right)\|z(k)\|.$$

Substituting the above function into (7.8) gives

$$V(k+1) - V(k) \leq -(\alpha_1 h - \beta_1 - \beta_2 \phi(u(k))) \frac{1}{h^2} V(k), \qquad (7.10)$$

where $\beta_2 = 2 \frac{\|P\|}{\lambda_{\min}(P)} (\|I\| + \|A - BK\|)(n + \|K\| \sqrt{n})$.

For the closed set Ω, there exists a constant r_1 such that $\Omega \subset \{x | \|x\| \leq r_1\}$. Then, since $\phi(u)$ is a continuous function, the constant $\bar{\phi}_\Omega$ is bounded on the closed set $\Omega_1 = \{u | |u| \leq r_1 \|K\| \frac{\lambda_{\max}(P)}{\lambda_{\min}(P)}\}$.

Let

$$h_\Omega = \max \left\{ \frac{2}{\alpha_1} (\beta_1 + \beta_2 \bar{\phi}_\Omega), 1 \right\}. \qquad (7.11)$$

It holds that $|u(0)| = |Kz(0)| \leq \|K\| \|z(0)\|$. Since $x(0) \in \Omega$ and $\|z(0)\| \leq \|x(0)\|$, one has $|u(0)| \leq r_1 \|K\|$, which means $u(0) \in \Omega_1$. Then, we have

$$V(1) - V(0) \leq -\frac{\alpha_1}{2h} V(0).$$

Assuming $k \leq m - 1$ with $m \geq 1$, it holds that

$$V(k+1) - V(k) \leq -\frac{\alpha_1}{2h} V(k).$$

Noticing that $V(m) \leq V(m-1) \leq \ldots \leq V(0)$, we have

$$\|z(m)\| \leq \frac{\lambda_{\max}(P)}{\lambda_{\min}(P)} \|z(0)\| \leq \frac{\lambda_{\max}(P)}{\lambda_{\min}(P)} |x(0)| \leq \frac{\lambda_{\max}(P)}{\lambda_{\min}(P)} r_1.$$

Thus, $|u(m)| \leq r_1 \|K\| \frac{\lambda_{\max}(P)}{\lambda_{\min}(P)}$ and $u(m) \in \Omega_1$, which yields $\phi(u(m)) \leq \bar{\phi}_\Omega$. From (7.10), we have

$$V(k+1) - V(k) \leq -\frac{\alpha_1}{2h} V(k)$$

for any $k \geq 0$. Moreover, we also have

$$V(k) \leq \left(1 - \frac{\alpha_1}{2h}\right)^k V(0).$$

Based on the above analysis, the convergence of $x(k)$ is obtained. This ends the proof. □

7.4 DESIGN OF DYNAMIC GAIN FEEDBACK CONTROL

In this section, a dynamic gain feedback strategy is presented to achieve global stabilization of the system (7.1). Consider the state transformation as

$$z_1(k) = \frac{x_1(k)}{h^n(k)}, z_2(k) = \frac{x_2(k)}{h^{n-1}(k)}, \ldots, z_n(k) = \frac{x_n(k)}{h(k)},$$

where $h(k)$ is the dynamic gain to be designed later.

Then, $z = (z_1, z_2, \ldots, z_n)^T$ satisfies

$$z_1(k+1) = l^n(k)z_1(k) + p_1 \frac{1}{h(k)} l^n(k)z_2(k) + l^n(k) \frac{1}{h^n(k)} f_1(x(k), u(k)),$$

$$z_2(k+1) = l^{n-1}(k)z_2(k) + p_2 \frac{1}{h(k)} l^{n-1}(k)z_3(k) + l^{n-1}(k) \frac{1}{h^{n-1}(k)} f_2(x(k), u(k)),$$

$$\vdots$$

$$z_n(k+1) = l(k)z_n(k) + p_n \frac{1}{h(k+1)} u(k) + l(k) \frac{1}{h(k)} f_n(x(k), u(k)),$$

where $l(k)$ is used to denote $\frac{h(k)}{h(k+1)}$.

Construct the dynamic gain feedback controller as:

$$u(k) = -Kz(k), \tag{7.12}$$

where K is the vector given in (7.3). The closed-loop system can be written as

$$z(k+1) = L(k)\Gamma z(k) + L(k)\tilde{F}(k),$$

where $L(k) = \text{diag}\left\{l^n(k), l^{n-1}(k), \ldots, l(k)\right\}$, $\Gamma = I + \frac{1}{h(k)}A - \frac{1}{h(k)}BK$ and

$$\tilde{F}(k) = \left(\frac{1}{h^n(k)} f_1(x(k), u(k)), \ldots, \frac{1}{h(k)} f_n(x(k), u(k))\right)^T.$$

Now, the main result of this section is summarized as follows:

Theorem 7.2

Suppose that Assumption 7.1 is satisfied. There exist $q_1 > 0$ and $q_2 > 0$ such that system (7.1) is globally stable under the control law (7.12) if the dynamic gain is designed as

$$h(k+1) = h(k) + \max\{q_1 \phi(u(k)) - h(k), 0\}, h(0) \geq q_2. \tag{7.13}$$

∎

Proof. From (7.13), it holds that

$$h(k+1) \geq h(k), \quad l(k) \leq 1,$$

and

$$h(k+1) \geq h(k) + q_1 \phi(u(k)) - h(k) \geq q_1 \phi(u(k)).$$

Consider the Lyapunov function candidate $V = z^T P z$. Then, we have

$$
\begin{aligned}
V(k+1) \\
= & z^T(k+1)Pz(k+1) \\
= & \left(L(k)\Gamma z(k)+L(k)\tilde{F}(k)\right)^T P\left(L(k)\Gamma z(k)+L(k)\tilde{F}(k)\right) \\
= & z^T(k)\Gamma^T L^T(k)PL(k)\Gamma z(k) \\
& + 2z^T(k)\Gamma^T L^T(k)PL(k)\tilde{F}(k)+\tilde{F}^T(k)L^T(k)PL(k)\tilde{F}(k).
\end{aligned}
$$

From Lemma 7.1, we have $L(k)^T PL(k) \leq l^{\alpha_2}(k)P$ due to $l(k) \leq 1$. Then, it holds that

$$
\begin{aligned}
& z^T(k)\Gamma^T L^T(k)PL(k)\Gamma z(k) \\
& \leq l^{\alpha_2}(k)z^T(k)\Gamma^T P\Gamma z(k) \\
& \leq l^{\alpha_2}(k)z^T(k)Pz(k)+\frac{l^{\alpha_2}(k)}{h(k)}z^T(k)\left((A-BK)^T P+P(A-BK)\right)z(k) \\
& \quad +\frac{l^{\alpha_2}(k)}{h^2(k)}z^T(k)(A-BK)^T P(A-BK)z(k) \\
& \leq \left(1-\frac{1}{h(k)}+\beta_1\frac{1}{h^2(k)}\right)l^{\alpha_2}(k)z^T(k)Pz(k),
\end{aligned}
$$

where β_1 satisfies $(A-BK)^T P(A-BK) \leq \beta_1 P$.

Since $h(k) \geq 1$, we have $1-\frac{1}{h(k)}+\frac{\beta_1}{h^2(k)} > 0$ and

$$
\left(1-\frac{1}{h(k)}+\beta_1\frac{1}{h^2(k)}\right)l^{\alpha_2}(k)z^T(k)Pz(k) \leq \left(1-\frac{1}{h(k)}+\beta_1\frac{1}{h^2(k)}\right)z^T(k)Pz(k).
$$

Meanwhile, similar to (7.9), the upper bounded of $\tilde{F}(k)$ is given as

$$
\left\|\tilde{F}(k)\right\| \leq \frac{1}{h^2(k)}\phi(u(k))\left(n+\|K\|\sqrt{n}\right)\|z(k)\|,
$$

which further gives

$$
\begin{aligned}
& 2z^T(k)\left(I+\frac{1}{h(k)}A-\frac{1}{h(k)}BK\right)^T L^T(k)PL(k)\tilde{F}(k) \\
& \leq 2\|z(k)\|\left\|I+\frac{1}{h(k)}A-\frac{1}{h(k)}BK\right\|\|L(k)\|^2\|P\|\left\|\tilde{F}(k)\right\| \\
& \leq \frac{\beta_2}{h^2(k)}l^2(k)\phi(u(k))z^T(k)Pz(k),
\end{aligned}
$$

where $\beta_2 = 2\|P\|\left(\|I\|+\|A-BK\|\right)\left(n+\|K\|\sqrt{n}\right)\frac{1}{\lambda_{\min}(P)}$.

Additionally, it holds that

$$\tilde{F}^T(k)L^T(k)PL(k)\tilde{F}(k) \leq \|P\| \, \|L(k)\|^2 \, \|\tilde{F}(k)\|^2$$

$$\leq \frac{\beta_3}{h^4(k)} l^2(k)\phi^2(u(k))z^T(k)Pz(k),$$

where $\beta_3 = (n + \|K\| \sqrt{n})^2 \|P\| \frac{1}{\lambda_{\min}(P)}$.
Then, we have

$$V(k+1) \leq \left(1 - \frac{1}{h(k)} + \frac{\beta_1}{h^2(k)}\right) V(k)$$

$$+ \frac{\beta_2}{h^2(k)} l^2(k)\phi(u(k))V(k) + \frac{\beta_3}{h^4(k)} l^2(k)\phi^2(u(k))V(k).$$

Choose

$$q_1 = 2\beta_2, \quad q_2 = \max\left\{4\left(\beta_1 + \frac{\beta_3}{4\beta_2^2}\right), 1\right\}.$$

Since $h(k+1) \geq 2\beta_2 \phi(u(k))$, we can get $1 \geq 2\frac{\beta_2}{h(k)} l(k)\phi(u(k))$. Thus, we have

$$V(k+1) \leq \left(1 - \frac{1}{2h(k)} + \left(\beta_1 + \beta_3 \frac{1}{4\beta_2^2}\right)\frac{1}{h^2(k)}\right)V(k).$$

Due to $h(k) \geq h(0) = 4\left(\beta_1 + \frac{\beta_3}{4\beta_2^2}\right)$, it holds that

$$V(k+1) \leq \left(1 - \frac{1}{4h(k)}\right)V(k)$$

which ensures the convergence of $V(k)$, i,e., $V(k) \leq V(0)$ holds for any k and $\lim_{k\to+\infty} V(k) = 0$. Meanwhile, since $z(k) \in \{z|z^T Pz \leq z^T(0)Pz(0)\}$ for any k, the boundness of $u(k)$ and $\phi(u(k))$ is ensured. Based on (7.2), $h(k)$ becomes a constant when $h(k) > 2\beta_2\phi(u(k))$. Therefore, the convergence of $x(k)$ is achieved, i.e., $\lim_{k\to+\infty}\|x(k)\| = 0$. □

7.5   SIMULATION

Now we present a numerical example to illustrate the effectiveness of the proposed control methods. Consider the system

$$x_1(k+1) = x_1(k) + x_2(k) - u(k)x_4(k) - u^2(k),$$
$$x_2(k+1) = x_2(k) + x_3(k) - u(k)x_4(k),$$
$$x_3(k+1) = x_3(k) + x_4(k) - u^2(k),$$
$$x_4(k+1) = x_4(k) + u(k),$$

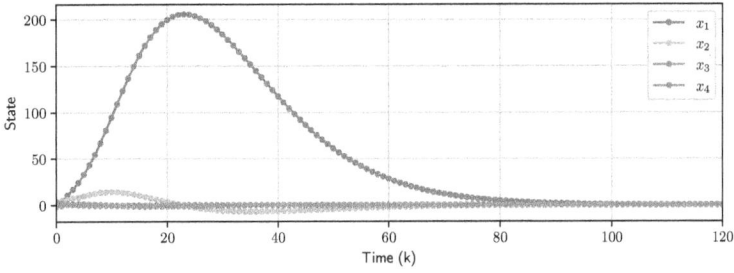

Figure 7.1 State trajectory.

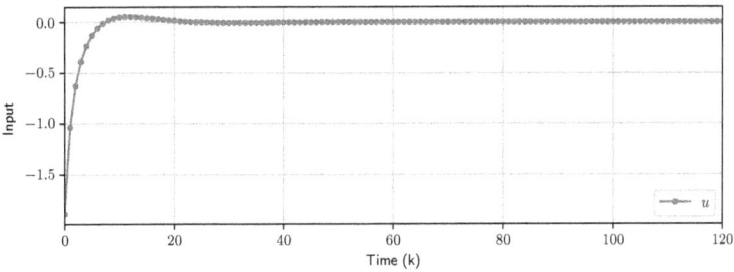

Figure 7.2 Input trajectory.

where $x = (x_1, x_2, x_3, x_4)^T \in \mathbb{R}^4$ is the system state, and $u \in \mathbb{R}$ is the input. It can be verified that Assumption 7.1 holds with $\phi(u) = u$.

I). Verification of Theorem 7.1: According to Theorem 7.1, the low-gain feedback controller is designed as

$$u(k) = -\frac{1}{h^4} x_1(k) - \frac{4}{h^3} x_2(k) - \frac{6}{h^2} x_3(k) - \frac{4}{h} x_4(k),$$

with the static gain $h = 8$. The simulation results are shown in Figures 7.1–7.2. It is shown that the system states converge to zero with the initial conditions given as $x(0) = (3, -2, -1, 4)^T$. This verifies the effectiveness of the proposed low-gain control approach.

II). Verification of Theorem 7.2: Based on Theorem 7.2, the dynamic gain feedback controller is designed as

$$u(k) = -\frac{1}{h^4(k)} x_1(k) - \frac{4}{h^3(k)} x_2(k) - \frac{6}{h^2(k)} x_3(k) - \frac{4}{h(k)} x_4(k),$$

$$h(k+1) = h(k) + \max\{1.2|u(k)| - h(k), 0\}$$

with $h(0) = 3$. The simulation results are shown in Figures 7.3–7.5. It can be observed that the system states converge to zero under the initial condition $x(0) = (3, -2, -1, 4)^T$, which verified our theoretical results.

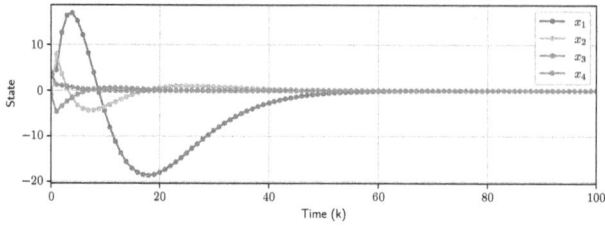

Figure 7.3 State trajectory.

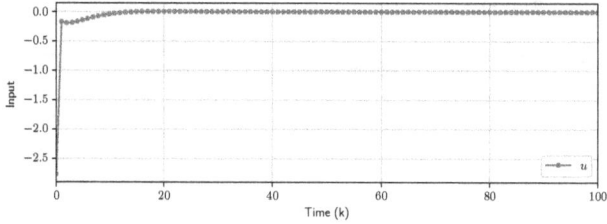

Figure 7.4 Input trajectory.

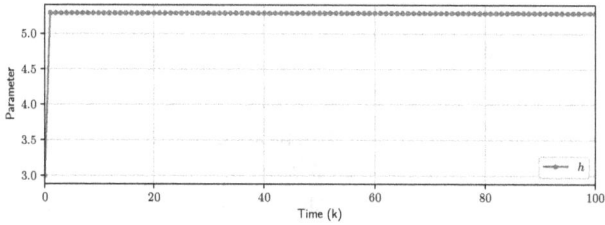

Figure 7.5 Trajectory of h.

7.6 NOTES

In this chapter, two static and dynamic gain control methods for the control of discrete-time nonlinear systems (DFNSs) have been investigated. It is assumed that the nonlinear growth rate of the considered DFNSs is characterized by an input-dependent function. The semi-global stabilization was ensured by using a static gain feedback control, and the condition for setting the gain is explicitly stated. Meanwhile, global stabilization has also been achieved by developing a dynamic gain control strategy. Simulation studies have verified the effectiveness of the proposed methods.

Part III

**Variable Gain Control for
Large-Scale Nonlinear Systems**

8 Variable Gain Control for Large-Scale Feedforward Nonlinear Systems

This chapter deals with the control of a class of large-scale interconnected nonlinear systems. First, we introduce a constant-gain decentralized control method in which each control input uses only the own states of the corresponding subsystem. With this method, semi-global stabilization of the concerned system is achieved. Next, we propose a centralized control method that achieves global stabilization. This method includes a dynamic gain designed with some global information, i.e., a control center which collects information from all subsystems is needed. Finally, we present a distributed control protocol to regulate the dynamic gain, which successfully eliminates the need for the above control center. The effectiveness of the proposed methods is illustrated by simulation results.

8.1 BACKGROUND

In recent decades, there has been an increased interest in the study of large-scale interconnected systems consisting of a large number of spatially distributed subsystems. Examples of such systems include power systems, water/gas distribution networks, transportation networks, multi-robot systems, wireless sensor networks, and unmanned factories [35, 17, 32]. An obvious feature of such systems is that the subsystems are closely interconnected (or coupled) with each other at the physical layer. To date, many remarkable works have been made on the control design of a wide variety of large-scale interconnected systems [56, 58, 110, 134, 31, 70, 156, 20].

Nonlinearities are common in practical dynamical systems. It has been acknowledged that proper gain scheduling provides an effective way to deal with the control problems caused by system nonlinearities [112, 160]. More specifically, gain scheduling allows the control gains to be adjusted online according to the ongoing system performance, thereby enabling the controller to better control the concerned system. For example, a dynamic high-gain scaling technique is proposed in [56, 58] to solve the output feedback control problem of a class of feedforward nonlinear systems. An adaptive dynamic gain scaling technique is proposed in [55] for a class of input-delayed systems in both feedforward and non-feedforward forms. Meanwhile, the high-gain technique is also used to solve the sampled-data control problem, the output feedback control problem, as well as the observer design problem for various nonlinear systems [22, 61, 126]. Recently, such a dynamic high-gain control approach has been generalized in [151, 155, 82, 153] to stabilize a class of large-scale nonlinear systems.

To ensure the stability of large-scale interconnected nonlinear systems, the convergence rates of all subsystems are required to be suitably regulated due to the nonlinear interactions (or couplings) among them. A typical method to meet the above requirement is to properly tune the control gains. For example, an intuitive solution is to deploy a global scheduling center which can regulate the gains of the local controllers, making all subsystems converge to the common equilibrium [151, 153, 155, 82]. Such a centralized gain scheduling method suffers from a number of inherent problems such as vulnerability to single-point failure and increased computation/communication burden, especially when the system scale gets larger.

This chapter will develop different control schemes for large-scale nonlinear interconnected systems. We first present a decentralized control method which does not need any communication support. However, this method can only achieve semi-global stabilization. Next, we propose a centralized control method which ensures global stabilization at the cost of a heavy communication burden. At last, we give a distributed control protocol which is able to achieve global stabilization with a reduced communication cost. Simulation results are provided to illustrate the effectiveness of the proposed methods.

8.2 PROBLEM DESCRIPTION

Consider the large-scale system

$$\dot{x}_i = A_i x_i + B_i u_i + \sum_{j=1}^{N} F_{i,j}(u_i, u_j, x_j), \quad i = 1, 2, \ldots, N, \tag{8.1}$$

where $x_i = (x_{i,1}, x_{i,2}, \ldots, x_{i,n_i})^T \in \mathbb{R}^{n_i}$ and $u_i \in \mathbb{R}$ represent the system state and input of the ith subsystem, respectively. The initial condition of the ith subsystem is defined as $x_i(t_0)$. A_i, B_i and $F_{i,j}$ are defined as

$$A_i = \begin{pmatrix} 0 & 1 & 0 & \cdots & 0 \\ 0 & 0 & 1 & \cdots & 0 \\ \vdots & \vdots & \vdots & & \vdots \\ 0 & 0 & 0 & \cdots & 1 \\ 0 & 0 & 0 & \cdots & 0 \end{pmatrix}, B_i = \begin{pmatrix} 0 \\ 0 \\ \vdots \\ 0 \\ 1 \end{pmatrix}, F_{i,j}(\cdot) = \begin{pmatrix} f_{i,j,1}(u_i, u_j, x_j) \\ f_{i,j,2}(u_i, u_j, x_j) \\ \vdots \\ f_{i,j,n_i-1}(u_i, u_j, x_j) \\ f_{i,j,n_i}(u_i, u_j, x_j) \end{pmatrix}.$$

The nonlinear term $f_{i,j,p}$ satisfies the following assumption.

Assumption 8.1. *For $i, j = 1, 2, \ldots, N$, the following inequalities hold*

$$\left| f_{i,j,p}(u_i, u_j, x) \right| \leq \varphi_i(u_i) \left(\sum_{q=\max\{2+n_j+p-n_i, 1\}}^{n_j+1} |x_q| + |u_j| \right), \quad p = 1, 2, \ldots, n_i - 1,$$

$$f_{i,j,n_i}(u_i, u_j, x, \tilde{x}) = 0$$

where $\varphi_i(u_i)$ is a continuous function with respect to u_i, and $x_{n_j+1} = 0$.

System (8.1) under Assumption 8.1 is a typical feedforward nonlinear system, and the control objective is to investigate different ways to stabilize this system.

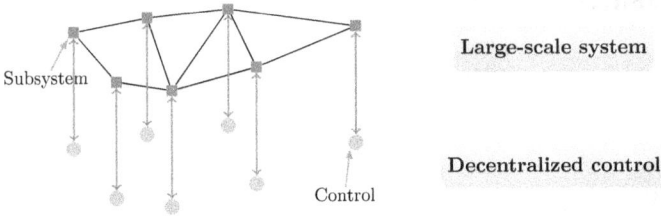

Figure 8.1 The framework of decentralized control.

8.3 DECENTRALIZED CONTROL WITH A CONSTANT GAIN

In this section, we design a decentralized control method with constant gains for system (8.1). The framework of this method is depicted in Figure 8.1.

8.3.1 CONTROL DESIGN

The decentralized control law is designed as

$$u_i = -\frac{k_{i,1}}{h^{n_i}}x_{i,1} - \frac{k_{i,2}}{h^{n_i-1}}x_{i,2} - \ldots - \frac{k_{i,n_i}}{h}x_{i,n_i}, \quad i = 1, 2, \ldots, N, \tag{8.2}$$

where $k_{i,1}, k_{i,2}, \ldots, k_{i,n_i}$ are constants making the polynomial

$$s^{n_i} + k_{i,1}s^{n_i-1} + \ldots + k_{i,n_i-1}s + k_{i,n_i}$$

Hurwitz, and $h \geq 1$ is the constant gain.

Consider the state transformation

$$z_{i,1} = \frac{x_{i,1}}{h^{n_i}}, z_{i,2} = \frac{x_{i,2}}{h^{n_i-1}}, \ldots, z_{i,n_i} = \frac{x_{i,n_i}v}{h}, \quad i = 1, 2, \ldots, N.$$

Then, we can get

$$\dot{z}_i = \frac{1}{h}\tilde{A}_i z_i + \sum_{j=1}^{N} \tilde{F}_{i,j}(u_i, u_j, x_j) \tag{8.3}$$

where

$$\tilde{A}_i = \begin{pmatrix} 0 & 1 & 0 & \cdots & 0 \\ 0 & 0 & 1 & \cdots & 0 \\ \vdots & \vdots & \vdots & & \vdots \\ 0 & 0 & 0 & \cdots & 1 \\ -k_{i,1} & -k_{i,2} & -k_{i,3} & \cdots & -k_{i,n_i} \end{pmatrix}, \quad \tilde{F}_{i,j}(\cdot) = \begin{pmatrix} \frac{1}{h^{n_i}}f_{i,j,1}(u_i, u_j, x_j) \\ \frac{1}{h^{n_i-1}}f_{i,j,2}(u_i, u_j, x_j) \\ \vdots \\ \frac{1}{h^2}f_{i,j,n_i-1}(u_i, u_j, x_j) \\ \frac{1}{h}f_{i,j,n_i}(u_i, u_j, x_j) \end{pmatrix}.$$

Since $\tilde{A}_i$ is Hurwitz, there exists a positive definite matrix P_i such that

$$\tilde{A}_i^T P_i + P_i \tilde{A}_i \leq -I.$$

8.3.2   STABILITY ANALYSIS

The main result of this section can be summarized as follows.

Theorem 8.1

Suppose that Assumption 8.1 is satisfied. For $x_i(0) \in \{x | \|x\| \leq \delta\}$, system (8.1) with the control law (8.2) is asymptotically stable if h satisfies

$$h \geq 2N \sum_{i=1}^{N} \left(n_i \|P_i\|^2 \bar{\varphi} + (\sqrt{n_i} + \|K_i\|)^2 \right) + 1,$$

where

$$\bar{\varphi} = \max_{i=1,2,\ldots,N} \max_{|q| \leq \bar{\delta}} \varphi_i^2(q)$$

and

$$\bar{\delta} = \max_{i=1,2,\ldots,N} \|K_i\| \sqrt{\frac{\lambda_{\max}(P_1) + \lambda_{\max}(P_2) + \ldots + \lambda_{\max}(P_N)}{\lambda_{\min}(P_i)}} \delta.$$

∎

Proof. Consider the Lyapunov function candidate

$$V = \sum_{i=1}^{N} z_i^T P_i z_i.$$

Its derivative is given as

$$\begin{aligned}
\dot{V}|_{(8.3)} &= \frac{1}{h} \sum_{i=1}^{N} z_i^T \left(\tilde{A}_i^T P_i + P_i \tilde{A}_i \right) z_i + 2 \sum_{i=1}^{N} \sum_{j=1}^{N} z_i^T P_i \tilde{F}_{i,j} \\
&\leq -\frac{1}{h} \sum_{i=1}^{N} \|z_i\|^2 + 2 \sum_{i=1}^{N} \sum_{j=1}^{N} z_i^T P_i \tilde{F}_{i,j}.
\end{aligned}$$ (8.4)

From Assumption 8.1, it holds that

$$\left| \frac{1}{h^{n_i+1-p}} f_{i,j,p}(u_i, u_j, x_j) \right|$$

$$\leq \varphi_i(u_i) \left(\sum_{q=\max\{2+n_j+p-n_i,1\}}^{n_j+1} \frac{1}{h^{n_i+1-p}} |x_{j,q}| + \frac{1}{h^{n_i+1-p}} |u_j| \right)$$

$$\leq \varphi_i(u_i) \left(\sum_{q=\max\{2+n_j+p-n_i,1\}}^{n_j+1} \frac{h^{n_j+1-q}}{h^{n_i+1-p}} |z_{j,q}| + \frac{1}{h^{n_i+1-p}} |K_j z_j| \right)$$

$$\leq \frac{1}{h^2} \varphi_i(u_i) \left(\sum_{q=\max\{2+n_j+p-n_i,1\}}^{n_j+1} |z_{j,q}| + |K_j z_j| \right)$$

$$\leq \frac{1}{h^2} \varphi_i(u_i) \left(\sqrt{n_j} + \|K_j\| \right) \|z_j\|$$

where $K_j = (k_{j,1}, k_{j,2}, \ldots, k_{j,n_j})$. Then, we get

$$\|\tilde{F}_{i,j}\| \leq \frac{1}{h^2} \varphi_i(u_i) \sqrt{n_i} \left(\sqrt{n_j} + \|K_j\| \right) \|z_j\|.$$

Back to (8.4), we have

$$\dot{V}|_{(8.3)} \leq -\frac{1}{h} \sum_{i=1}^{N} \|z_i\|^2 + 2 \sum_{i=1}^{N} \sum_{j=1}^{N} \frac{1}{h^2} \varphi_i(u_i) \|P_i\| \sqrt{n_i} \left(\sqrt{n_j} + \|K_j\| \right) \|z_j\| \|z_i\|$$

$$\leq -\frac{1}{h} \sum_{i=1}^{N} \|z_i\|^2 + \sum_{i=1}^{N} \sum_{j=1}^{N} \frac{1}{h^2} \left(n_i \|P_i\|^2 \varphi_i^2(u_i) \|z_i\|^2 + \left(\sqrt{n_j} + \|K_j\| \right)^2 \|z_j\|^2 \right)$$

$$\leq -\frac{1}{h} \sum_{i=1}^{N} \|z_i\|^2 + N \frac{1}{h^2} \sum_{i=1}^{N} \left(n_i \|P_i\|^2 \varphi_i^2(u_i) + \left(\sqrt{n_i} + \|K_i\| \right)^2 \right) \|z_i\|^2.$$

Next, we use mathematical induction to prove $\dot{V} \leq -\frac{1}{2h} \sum_{i=1}^{N} \|z_i(t)\|^2$. With the design of h, it holds that $\dot{V}(t_0) < -\frac{1}{2h} \sum_{i=1}^{N} \|z_i(t_0)\|^2$.

Inductive step: Suppose $\dot{V}(t) < -\frac{1}{2h} \sum_{i=1}^{N} \|z_i(t)\|^2$ holds during the period $t \in [t_0, t_1)$, then we can get $V(t_1) \leq V(t_0)$. From the definition of u_i we have

$$|u_i(t)| \leq \|K_i\| \|z_i(t)\| \leq \|K_i\| \sqrt{\frac{V(t_0)}{\lambda_{\min}(P_i)}}$$

$$\leq \|K_i\| \sqrt{\frac{\lambda_{\max}(P_1) + \lambda_{\max}(P_2) + \ldots + \lambda_{\max}(P_N)}{\lambda_{\min}(P_i)}} \delta \leq \bar{\delta},$$

which can further yield

$$\dot{V}(t_1) < -\frac{1}{2h} \sum_{i=1}^{N} \|z_i(t_1)\|^2.$$

Using the inductive argument above, one concludes that

$$\dot{V}(t) \le -\frac{1}{2h}\sum_{i=1}^{N}\|z_i(t)\|^2$$

for any $t \in [t_0, +\infty)$. Since h is a constant, we have that the closed-loop system consisting of (8.1) and (8.2) is stable. This completes the proof. □

8.3.3 SIMULATION

In this section, we verify the effectiveness of the above decentralized control method by considering the following nonlinear system

$$
\begin{cases}
\dot{x}_{1,1} = x_{1,2} - 2u_1 x_{1,3} + u_1 x_{2,3} + u_1 x_{3,3}, \\
\dot{x}_{1,2} = x_{1,3} - u_1^2, \\
\dot{x}_{1,3} = u_1,
\end{cases}
$$
$$
\begin{cases}
\dot{x}_{2,1} = x_{2,2} + 3u_2 x_{1,3} - u_2 x_{2,3} + u_2 x_{3,3}, \\
\dot{x}_{2,2} = x_{2,3} - u_2^2, \\
\dot{x}_{2,3} = u_2,
\end{cases}
\tag{8.5}
$$
$$
\begin{cases}
\dot{x}_{3,1} = x_{3,2} + 2u_3 x_{1,3} + u_3 x_{2,3} - u_3 x_{3,3}, \\
\dot{x}_{3,2} = x_{3,3} + u_3^2, \\
\dot{x}_{3,3} = u_3,
\end{cases}
$$

where $x_i = (x_{i,1}, x_{i,2}, x_{i,3})^T \in \mathbb{R}^3$ is the state, $u_i \in \mathbb{R}$ is the input, and $y_i \in \mathbb{R}$ is the output of the ith subsystem, $i = 1, 2, 3$. It can be verified that the nonlinear terms satisfy Assumption 8.1.

The local decentralized control is designed as

$$u_i = -\frac{1}{h^3}x_{i,1} - \frac{3}{h^2}x_{i,2} - \frac{3}{h}x_{i,2}, \quad i = 1, 2, 3,$$

with $h = 10$. The simulation results are given in Figure 8.2, when the initial condition is $x_1(0) = (1.25, 1.02, 1.71)^T$, $x_2(0) = (6.00, -0.12, -0.99)^T$, $x_3(0) = (0.46, -0.11, -5.93)^T$. It can be observed that the system states all converge to zero, which is consistent without theoretical analyses.

8.4 CENTRALIZED CONTROL WITH A GLOBAL DYNAMIC GAIN

It should be noted that the above method can only achieve semi-global stabilization. In this section, we consider a modified control method with a global dynamic gain to achieve global stabilization. The framework of this method is depicted in Figure 8.3.

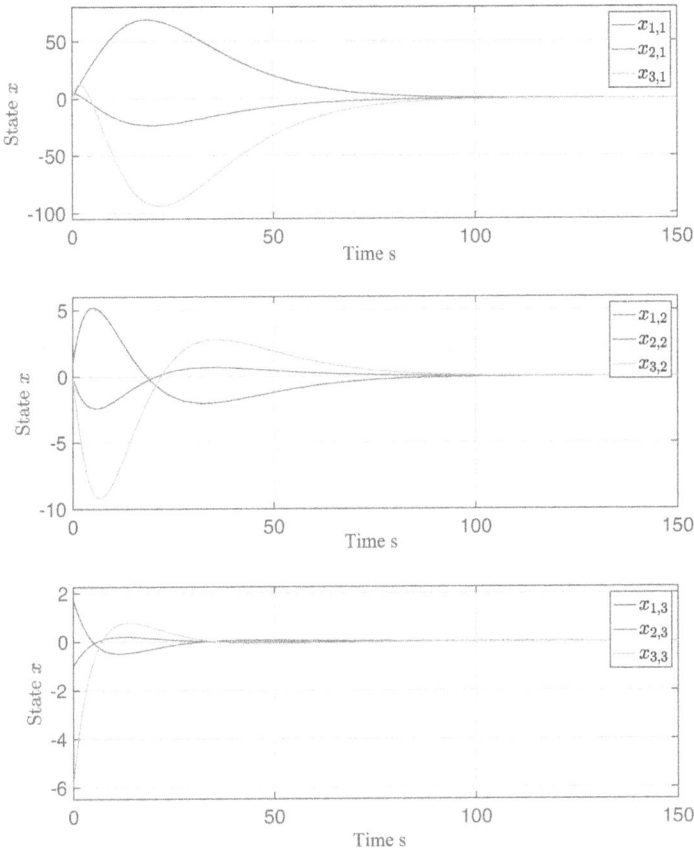

Figure 8.2 The control performance.

8.4.1 CONTROL DESIGN

The control law is modified as

$$u_i = -\frac{k_{i,1}}{h^{n_i}}x_{i,1} - \frac{k_{i,2}}{h^{n_i-1}}x_{i,2} - \ldots - \frac{k_{i,n_i}}{h}x_{i,n_i}, \quad i = 1,2,\ldots,N, \tag{8.6}$$

where $k_{i,1}, k_{i,2}, \ldots, k_{i,n_i}$ are constants, and $h \geq 1$ is the dynamic gain.

Consider the state transformation

$$z_{i,1} = \frac{x_{i,1}}{h^{n_i}}, z_{i,2} = \frac{x_{i,2}}{h^{n_i-1}}, \ldots z_{i,n_i} = \frac{x_{i,n_i}}{h}, \quad i = 1,2,\ldots,N.$$

Then, we can get

$$\dot{z}_i = \frac{1}{h}\tilde{A}_i z_i + \sum_{j=1}^{N} \tilde{F}_{i,j}(u_i, u_j, x_j) - \frac{\dot{h}}{h}D_i z_i, \tag{8.7}$$

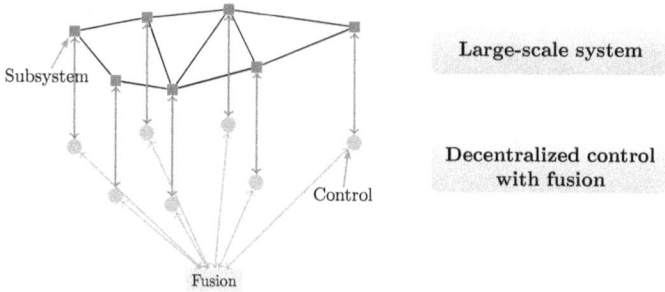

Figure 8.3 The framework of decentralized control with global dynamic parameters.

where

$$
\tilde{A}_i =
\begin{pmatrix}
0 & 1 & 0 & \cdots & 0 \\
0 & 0 & 1 & \cdots & 0 \\
\vdots & \vdots & \vdots & & \vdots \\
0 & 0 & 0 & \cdots & 1 \\
-k_{i,1} & -k_{i,2} & -k_{i,3} & \cdots & -k_{i,n_i}
\end{pmatrix},
\quad
\tilde{F}_{i,j}(\cdot) =
\begin{pmatrix}
\frac{1}{h^{n_i}} f_{i,j,1}(u_i,u_j,x_j) \\
\frac{1}{h^{n_i-1}} f_{i,j,2}(u_i,u_j,x_j) \\
\vdots \\
\frac{1}{h^2} f_{i,j,n_{n_i}-1}(u_i,u_j,x_j) \\
\frac{1}{h} f_{i,j,n_{n_i}}(u_i,u_j,x_j)
\end{pmatrix}
$$

and $D_i = \mathrm{diag}\{n_i, n_i - 1, \ldots, 1\}$.

We can find a vector $K_i = (k_{i,1}, k_{i,2}, \ldots, k_{i,n_i})$ and the matrix P_i such that

$$
\tilde{A}_i^T P_i + P_i \tilde{A}_i \le -I, \quad P_i D_i + D_i P_i \ge \alpha_i I
$$

hold with α_i being a positive constant.

Based on K_i and P_i, h is designed as

$$
\dot{h} = \frac{1}{\alpha h} \max\{g(u_1, u_2, \ldots, u_N) - \frac{h}{2}, 0\}, \quad h(t_0) = 1, \tag{8.8}
$$

where $\alpha = \max\{\alpha_1, \alpha_2, \ldots, \alpha_N\}$, and

$$
g(u_1, u_2, \ldots, u_N) = N \max_{i=1,2,\ldots,N} \left\{ n_i \|P_i\|^2 \varphi_i^2(u_i) + (\sqrt{n_i} + \|K_i\|)^2 \right\}.
$$

8.4.2 STABILITY ANALYSIS

We summarize the main result as follows:

Theorem 8.2

Suppose that Assumption 8.1 is satisfied. System (8.1) is globally stable with the control law composed of (8.6) and (8.8). ∎

Proof. For the closed-loop system (8.7), we consider the Lyapunov function candidate

$$V = \sum_{i=1}^{N} z_i^T P_i z_i.$$

Then, its derivative is computed as

$$\dot{V}|_{(8.7)} = \frac{1}{h} \sum_{i=1}^{N} z_i^T \left(\tilde{A}_i^T P_i + P_i \tilde{A}_i \right) z_i + 2 \sum_{i=1}^{N} \sum_{j=1}^{N} z_i^T P_i \tilde{F}_{i,j} - \frac{\dot{h}}{h} \sum_{i=1}^{N} z_i^T \left(D_i P_i + P_i D_i \right) z_i$$

$$\leq -\frac{1}{h} \sum_{i=1}^{N} \|z_i\|^2 + 2 \sum_{i=1}^{N} \sum_{j=1}^{N} z_i^T P_i \tilde{F}_{i,j} - \alpha_i \frac{\dot{h}}{h} \sum_{i=1}^{N} \|z_i\|^2.$$

For the nonlinear term $\tilde{F}_{i,j}$, we can also deduce from Assumption 8.1 that

$$\|\tilde{F}_{i,j}\| \leq \frac{1}{h^2} \varphi_i(u_i) \sqrt{n_i} \left(\sqrt{n_j} + \|K_j\| \right) \|z_j\|.$$

Back to (8.4), we have

$$\dot{V}|_{(8.7)} \leq -\frac{1}{h} \sum_{i=1}^{N} \|z_i\|^2 - \alpha_i \frac{\dot{h}}{h} \sum_{i=1}^{N} \|z_i\|^2$$

$$+ 2 \sum_{i=1}^{N} \sum_{j=1}^{N} \frac{1}{h^2} \varphi_i(u_i) \|P_i\| \sqrt{n_i} \left(\sqrt{n_j} + \|K_j\| \right) \|z_j\| \|z_i\|$$

$$\leq -\frac{1}{h} \sum_{i=1}^{N} \|z_i\|^2 - \alpha_i \frac{\dot{h}}{h} \sum_{i=1}^{N} \|z_i\|^2$$

$$+ N \frac{1}{h^2} \sum_{i=1}^{N} \left(n_i \|P_i\|^2 \varphi_i^2(u_i) + \left(\sqrt{n_i} + \|K_i\| \right)^2 \right) \|z_i\|^2$$

$$\leq -\frac{1}{h} \sum_{i=1}^{N} \|z_i\|^2 + \frac{1}{h^2} \sum_{i=1}^{N} g(u_1, u_2, \ldots, u_N) \|z_i\|^2 - \alpha \frac{\dot{h}}{h} \sum_{i=1}^{N} \|z_i\|^2$$

$$\leq -\frac{1}{2h} \sum_{i=1}^{N} \|z_i\|^2.$$

Thus, the closed-loop system (8.7) is globally stable. Since

$$|u_i(t)| \leq \|K_i\| \|z_i(t)\| \leq \|K_i\| \sqrt{\frac{V(t_0)}{\lambda_{\min}(P_i)}},$$

we can get that $g(u_1, u_2, \ldots, u_N)$ are bounded, which further leads to the boundedness of h. Therefore, the stabilization of system (8.7) is guaranteed. This ends the proof. $\square$

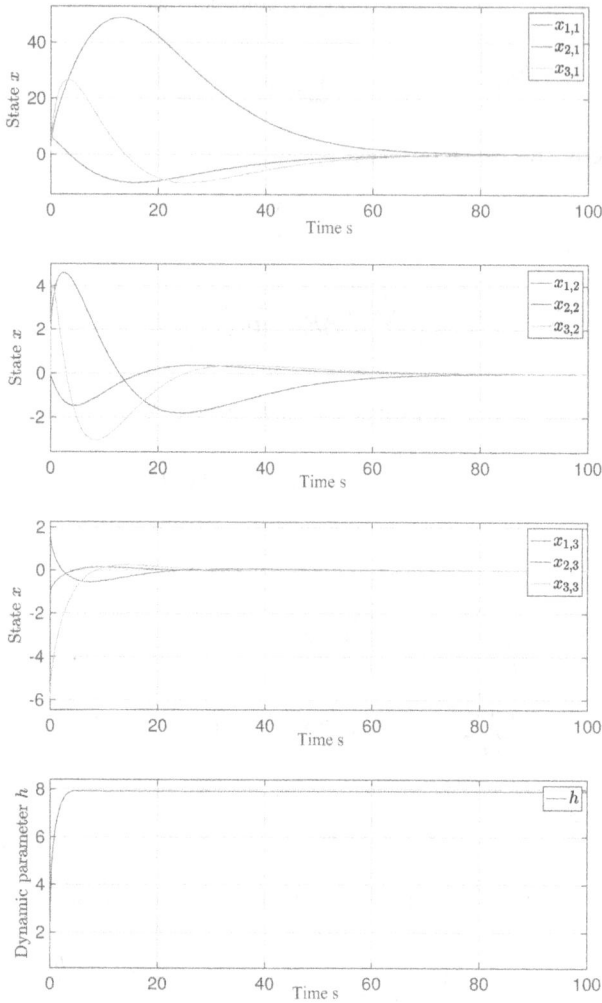

Figure 8.4 The control performance.

8.4.3 SIMULATION

Consider system (8.5) and design the controller as

$$u_i = -\frac{1}{h^3}x_{i,1} - \frac{3}{h^2}x_{i,2} - \frac{3}{h}x_{i,2}, \quad i = 1,2,3,$$

where h satisfies

$$\dot{h} = \frac{1}{h}\max\{5(|u_1| + |u_2| + u_2^2 + |u_3|) + 1 - \frac{h}{2}, 0\}, \quad h(t_0) = 1.$$

It can be seen from Figure 8.4 that all the system state converges towards zero with a faster speed than the decentralized method introduced above. This shows the advantage of the centralized method over the decentralized one.

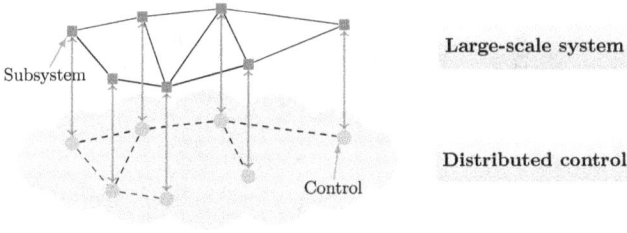

Figure 8.5 The framework of distributed control.

8.5 CONTROL WITH DISTRIBUTED DYNAMIC GAINS

Notice that the design of h in the above method requires the input signals of all subsystems to be available. To remove this requirement, we propose a decentralized control method with distributed dynamic gains. The framework is depicted in Figure 8.5.

Before the control design, we introduce the basic concepts of graph theory, which is used to describe the communication network.

The communication network topology of N nodes is modeled by a directed graph $\mathcal{G} = \{\mathcal{V}, \mathcal{E}, \mathcal{A}\}$, where $\mathcal{V} = \{1, 2, \ldots, N\}$ denotes a node set, $\mathcal{E}$ denotes the set of paired edges, and $\mathcal{A} = (a_{ij})_{N \times N}$ represents the weight matrix with $a_{ij} > 0$ denoting the weight of the edge from node j to node i. For any two distinct nodes $i, j \in \mathcal{V}$, we have $(i, j) \in \mathcal{E}$ if there is a directed edge from node j to node i, and $(i, j) \notin \mathcal{E}$ otherwise. We assume that there is no self-loop, i.e., $(i, i) \notin \mathcal{E}$ and $a_{ii} = 0$. The set of neighbors of node $i \in \mathcal{V}$ is denoted by $\mathcal{N}_i = \{j \in \mathcal{V} : (i, j) \in \mathcal{E}\}$. Denote by m_i the number of neighbors of node i (or the cardinality of $\mathcal{N}_i$). A directed path (i_1, i_j) in $\mathcal{G}$ is a finite sequence of directed edges $(i_1, i_2), \ldots, (i_{j-1}, i_j)$. In the following, it is assumed that $\mathcal{G}$ is strongly connected, which means that for any two distinct nodes $i, j \in \mathcal{V}$, there is a directed path from node j to node i.

8.5.1 CONTROL DESIGN

The control law is designed as

$$u_i = -\frac{k_{i,1}}{h_i^{n_i}} x_{i,1} - \frac{k_{i,2}}{h_i^{n_i-1}} x_{i,2} - \ldots - \frac{k_{i,n_i}}{h_i} x_{i,n_i}, \quad i = 1, 2, \ldots, N, \quad (8.9)$$

where $k_{i,1}$ to k_{i,n_i} are constants to be chosen, and $h_i \geq 1$ is the distributed dynamic gain.

By conducting the state transformation

$$z_{i,1} = \frac{x_{i,1}}{h_i^{n_i}}, z_{i,2} = \frac{x_{i,2}}{h_i^{n_i-1}}, \ldots z_{i,n_i} = \frac{x_{i,n_i}}{h_i}, \quad i = 1, 2, \ldots, N,$$

yields

$$\dot{z}_i = \frac{1}{h_i}\tilde{A}_i z_i + \sum_{j=1}^{N} \tilde{F}_{i,j}(u_i, u_j, x_j) - \frac{\dot{h}_i}{h_i} D_i z_i, \tag{8.10}$$

where

$$\tilde{A}_i = \begin{pmatrix} 0 & 1 & 0 & \cdots & 0 \\ 0 & 0 & 1 & \cdots & 0 \\ \vdots & \vdots & \vdots & & \vdots \\ 0 & 0 & 0 & \cdots & 1 \\ -k_{i,1} & -k_{i,2} & -k_{i,3} & \cdots & -k_{i,n_i} \end{pmatrix}, \quad \tilde{F}_{i,j}(\cdot) = \begin{pmatrix} \frac{1}{h^{n_i}} f_{i,j,1}(u_i, u_j, x_j) \\ \frac{1}{h^{n_i-1}} f_{i,j,2}(u_i, u_j, x_j) \\ \vdots \\ \frac{1}{h^2} f_{i,j,n_i-1}(u_i, u_j, x_j) \\ \frac{1}{h} f_{i,j,n_i}(u_i, u_j, x_j) \end{pmatrix}$$

and $D_i = \mathrm{diag}\{n_i, n_i - 1, \ldots, 1\}$.

We can find a vector $K_i = (k_{i,1}, k_{i,2}, \ldots, k_{i,n_i})$ and a positive definite matrix P_i such that

$$\tilde{A}_i^T P_i + P_i \tilde{A}_i \leq -I, \quad P_i D_i + D_i P_i \geq \tau_i I$$

with τ_i being a positive constant.

Now, we give the following lemma.

Lemma 8.1

For system (8.1) under Assumption 8.1 and the local control law (8.9), if h_i satisfies

$$\dot{h}_i \geq \max\left\{\frac{1}{h_i}(\rho_i(u_i) + \alpha_i) - \frac{1}{2\tau_i}, \ 0\right\}, \quad h_i(t_0) \geq 1$$

where $\rho_i(u_i) = n_i N \|P_i\| \varphi_i^2(u_i)\frac{1}{\tau_i}$ and $\alpha_i > 0$, then there exist positive scalars κ_1 and κ_2 such that the function $V_i = z_i^T P_i z_i$ satisfies

$$\dot{V}_i|_{(8.10)} \leq -\frac{\kappa_1}{h_i}\left(1 + \frac{\alpha_i}{h_i}\right) V_i + \sum_{j=1}^{N} \frac{\kappa_2}{h_j^2}\left(1 + \frac{h_j}{h_i}\right)^{2n} V_j. \tag{8.11}$$

∎

Proof. The derivative of V_i is computed as

$$\dot{V}_i|_{(8.10)} = \frac{1}{h_i} z_i^T \left(\tilde{A}_i^T P_i + P_i \tilde{A}_i\right) z_i + 2\sum_{j=1}^{N} z_i^T P_i \tilde{F}_{i,j} - \frac{\dot{h}_i}{h_i} z_i^T \left(D_i P_i + P_i D_i\right) z_i$$

$$\leq -\frac{1}{h_i}\|z_i\|^2 + 2\sum_{j=1}^{N} z_i^T P_i \tilde{F}_{i,j} - \tau_i \frac{\dot{h}_i}{h_i}\|z_i\|^2. \tag{8.12}$$

From Assumption 8.1, one obtains

$$\left| \frac{1}{h_i^{n_i+1-p}} f_{i,j,p}(u_i, u_j, x_j) \right|$$

$$\leq \varphi_i(u_i) \frac{h_j^{n_i+1-p}}{h_i^{n_i+1-p}} \left(\sum_{q=\max\{2+n_j+p-n_i,1\}}^{n_j+1} \frac{1}{h_j^{n_i+1-p}} |x_{j,q}| + \frac{1}{h_j^{n_i+1-p}} |u_j| \right)$$

$$\leq \varphi_i(u_i) \frac{h_j^{n_i+1-p}}{h_i^{n_i+1-p}} \left(\sum_{q=\max\{2+n_j+p-n_i,1\}}^{n_j+1} \frac{h_j^{n_j+1-q}}{h_j^{n_i+1-p}} |z_{j,q}| + \frac{1}{h_j^{n_i+1-p}} |K_j z_j| \right)$$

$$\leq \frac{1}{h_j^2} \frac{h_j^{n_i+1-p}}{h_i^{n_i+1-p}} \varphi_i(u_i) \left(\sum_{q=\max\{2+n_j+p-n_i,1\}}^{n_j+1} |z_{j,q}| + |K_j z_j| \right)$$

$$\leq \frac{1}{h_j^2} \frac{h_j^{n_i+1-p}}{h_i^{n_i+1-p}} \varphi_i(u_i) \left(\sqrt{n_j} + \|K_j\| \right) \|z_j\|,$$

where $K_j = (k_{j,1}, k_{j,2}, \ldots, k_{j,n_j})$. Since

$$\frac{1}{h_j^2} \frac{h_j^{n_i+1-p}}{h_i^{n_i+1-p}} \leq \frac{1}{h_i h_j} \frac{h_j^{n_i-p}}{h_i^{n_i-p}} \leq \frac{1}{h_i h_j} \left(1 + \frac{h_j}{h_i} \right)^{n_i},$$

we get

$$\|\tilde{F}_{i,j}\| \leq \frac{1}{h_i h_j} \left(1 + \frac{h_j}{h_i} \right)^{n_i} \varphi_i(u_i) \sqrt{n_i} \left(\sqrt{n_j} + \|K_j\| \right) \|z_j\|.$$

Back to (8.12), we have

$$\dot{V}_i|_{(8.10)} \leq -\frac{1}{h_i} \|z_i\|^2 - \tau_i \frac{\dot{h}_i}{h_i} \|z_i\|^2$$

$$+ 2 \sum_{j=1}^{N} \frac{1}{h_i h_j} \left(1 + \frac{h_j}{h_i} \right)^{n_i} \varphi_i(u_i) \|P_i\| \sqrt{n_i} \left(\sqrt{n_j} + \|K_j\| \right) \|z_j\| \|z_i\|$$

$$\leq -\frac{1}{h_i} \|z_i\|^2 + \frac{n_i N \|P_i\|^2}{h_i^2} \varphi_i^2(u_i) \|z_i\|^2$$

$$+ \sum_{j=1}^{N} \left(\sqrt{n_j} + \|K_j\| \right)^2 \frac{1}{h_j^2} \left(1 + \frac{h_j}{h_i} \right)^{2n} \|z_j\|^2 - \tau_i \frac{\dot{h}_i}{h_i} \|z_i\|^2$$

$$\leq -\frac{1}{2h_i} \left(1 + \frac{\alpha_i}{h_i} \right) \|z_i\|^2 + + \sum_{j=1}^{N} \left(\sqrt{n_j} + \|K_j\| \right)^2 \frac{1}{h_j^2} \left(1 + \frac{h_j}{h_i} \right)^{2n} \|z_j\|^2.$$

Let $\kappa_1 = \frac{n_i}{2\lambda_{\max}(P_i)}$ and $\kappa_2 = \left(\sqrt{n_j} + \|K_j\| \right)^2 \frac{1}{\lambda_{\min}(P)}$, then (8.11) holds. This ends the proof. $\square$

Under a strongly connected network $\mathcal{G}$, we design h_i as

$$
\dot{h}_i = \max\left\{ \frac{\rho_i(u_i)+\alpha_i}{h_i} - \frac{1}{2\tau_i}, -a_{i1}\frac{h_i-h_1}{\beta+(h_i-h_1)}, \ldots, -a_{iN}\frac{h_i-h_N}{\beta+h_i-h_N}, 0 \right\}, \quad (8.13)
$$
$$
h_i(t_0) \in [1,\beta],
$$

where a_{ij} is the weight of the edge from controller j to controller i, $\forall\, j \in \{1,2,\ldots,N\}$, $\rho_i(u_i)$ and τ_i is provided in Lemma 8.1, α_i and β are design parameters.

8.5.2 STABILITY ANALYSIS

We summarize the main result as follows:

Theorem 8.3

Suppose that Assumption 8.1 is satisfied. System (8.1) can be globally stabilized through the control law (8.9) if h_i is designed as (8.13) and α_i satisfies

$$
\alpha_i \geq \frac{\kappa_2}{\kappa_1}N(2+(N-1)\beta)^{2n}.
$$

■

Proof. Consider the closed-loop system (8.10). Let

$$
V = \sum_{i=1}^{N} V_i,
$$

with $V_i = z_i^T P_i z_i$ being provided in Lemma 8.1. Then, we get

$$
\dot{V}|_{(8.10)} \leq -\sum_{i=1}^{N} \frac{\kappa_1}{h_i}\left(1+\frac{\alpha_i}{h_i}\right)V_i + \sum_{i=1}^{N}\sum_{j=1}^{N} \frac{\kappa_2}{h_j^2}\left(1+\frac{h_j}{h_i}\right)^{2n}V_j
$$
$$
\leq -\sum_{i=1}^{N} \frac{\kappa_1}{h_i}V_i - \sum_{i=1}^{N}\sum_{j=1}^{N} \frac{\kappa_2}{h_j^2}\left((2+(N-1)\beta)^{2n}-\left(1+\frac{h_j}{h_i}\right)^{2n}\right)V_j.
$$

The rest of the proof is divided into three steps. First, we consider the boundedness of $\rho_i(u_i), i = 1,2,\ldots,N$, under the condition $h_j - h_i \leq \beta$, $a_{ij} \neq 0$. Then, we prove that $h_j - h_i \leq \beta$, $a_{ij} \neq 0$ always holds under (8.13). At last, we prove that the max-consensus of h_i is achieved, which guarantees the asymptotic stability of the closed-loop system at the equilibrium point $x_i = 0$.

Step 1: Proof of the boundedness of $\rho_i(u_i)$ when $h_j - h_i \leq \beta$, $a_{ij} \neq 0$.
From (8.13), it holds that $h_i(t) \geq h_i(0) \geq 1$ due to $\dot{h}_i(t) \geq 0$.

If $h_j \leq h_i$, one has $\frac{h_j}{h_i} \leq 1$. Now we consider the case $h_j > h_i$. For a strongly connected digraph $\mathscr{G}$, there always exists a path $(i_1,i_2),(i_2,i_3),\ldots,(i_{p-1},i_p) \in \mathscr{E}$ linking any two distinct nodes i and j with $i = i_1$ and $j = i_p$. For the edge $(i_k,i_{k+1}), k = 1,2,\ldots,p-1$, it holds that $a_{i_k i_{k+1}} > 0$ and $h_{i_{k+1}} - h_{i_k} \leq \beta$. Meanwhile, the number of these edges must be less than $N - 1$. Thus, when $h_j > h_i$, one has

$$h_j - h_i = (h_{i_1} - h_{i_2}) + \ldots + (h_{i_{k-1}} - h_{i_k}) \leq (N-1)\beta,$$

and

$$\frac{h_j}{h_i} \leq 1 + \frac{1}{h_i}(N-1)\beta \leq 1 + (N-1)\beta.$$

Therefore, we conclude that $h_j/h_i \leq (N-1)\beta + 1, i,j \in \{1,2,\ldots,N\}$. Now, we can get

$$\dot{V}|_{(8.10)} \leq -\frac{\kappa_1}{h}V. \tag{8.14}$$

In view of the Lyapunov theory, we conclude that when $h_j - h_i \leq \beta$, $a_{ij} \neq 0$, $V(t)$ exists and $V(t) \leq V(t_0)$ holds. With $u_i = K_i z_i$, we have that $|u_i| \leq \|K_i\|\sqrt{\frac{\lambda_{\min}(P_i)}{V(t_0)}}$ with $\lambda_{\min}(P_i)$ being the minimum eigenvalue of matrix P_i. Since $\rho_i(u_i)$ is a continuous function, one can always find a scalar M such that $\rho_i(u_i) \leq M$ for any $i = 1,2,\ldots,N$ if $h_j - h_i \leq \beta$, $a_{ij} \neq 0$ holds.

Step 2: Proof of $h_j(t) - h_i(t) \leq \beta$ when $a_{ij} \neq 0$ holds for all t and $i,j \in \{1,2,\ldots,N\}$.

We prove this by contradiction. Suppose that there is an instant $s_0 > 0$ such that $h_j(s_0) - h_i(s_0) > \beta$, $a_{ij} \neq 0$. It is noted that the initial condition satisfies $h_j(t_0) - h_i(t_0) \leq \beta - 1 < \beta$, $a_{ij} \neq 0$. Since $\dot{h}_j(t) - \dot{h}_i(t)$ exists and $h_j(t) - h_i(t)$ is continuous, there is a finite time instant $s_1 \in [t_0, s_0)$ such that $h_j(t) - h_i(t) < \beta$, $a_{ij} \neq 0$ in $[t_0, s_1)$ and $h_j(s_1) - h_i(s_1) = \beta$, $a_{ij} \neq 0$.

Notice that when $h_j(s_1) - h_i(s_1) = \beta$, $a_{ij} \neq 0$, we have that

$$\lim_{t \to s_1} \dot{h}_i(t) \geq \lim_{t \to s_1} -a_{ij}\frac{h_i(t) - h_j(t)}{\beta + h_i(t) - h_j(t)} = +\infty,$$

which means there always exists at least one h_i satisfying $\lim_{t \to s_1} \dot{h}_i(t) = +\infty$. Now, we consider h_{i_1} satisfying $\lim_{t \to s_1} \dot{h}_{i_1}(t) = +\infty$ and $h_{i_1}(s_1) \geq h_j(s_1)$ for h_j satisfying $\lim_{t \to s_1} \dot{h}_j(t) = +\infty$.

During the period $t \in [t_0, s_1]$, we can deduce the boundness of h_i from the condition $\rho_i(u_i(t)) \leq M, t \in [t_0, s_1]$. This is because $h_i(t)$ cannot be larger than $\max_{i \in \{1,2,\ldots,N\}}\{2\tau_i(M + \alpha_i)\}$. Consider the set $\mathscr{Q} = \{i|a_{i_1 i} \neq 0, h_i(s_1) \geq h_{i_1}(s_1), i = 1,2,\ldots,N\}$, which must be non-empty. Otherwise, we have $-a_{i_1 j}\frac{h_{i_1}(s_1) - h_j(s_1)}{\beta + h_{i_1}(s_1) - h_j(s_1)} \leq 0$ for all $i = 1,2,\ldots,N$, and $\dot{h}_{i_1}(s_1) \leq M + \alpha_i$. This contradicts the choice of i_1.

Furthermore, from the definition of $\mathscr{Q}$, there exists a constant M_1 such that $\dot{h}_i(t) \leq M_1, t \in [t_0, s_1]$, for any $i \in \mathscr{Q}$. $\lim_{t \to s_1} \dot{h}_{i_1}(t) = +\infty$ indicates that there exists an instant $s_2 \in (t_0, s_1)$ such that $\dot{h}_{i_1}(t) \geq M_1, t \in [s_2, s_1)$. Then, we have that, for $t \in [s_2, s_1), i \in$

$\mathscr{Q}$, the condition $\dot{h}_i(t) - \dot{h}_{i_1}(t) \le 0$ holds, which means that $h_i(t) - h_{i_1}(t) \le h_i(s_2) - h_{i_1}(s_2) \le \beta$. Thus, it holds that

$$-a_{i_1 i} \frac{h_{i_1}(t) - h_i(t)}{\beta + h_{i_1}(t) - h_i(t)} \le a_{i_1 i} \frac{\beta}{\beta + h_{i_1}(s_2) - h_i(s_2)}, \quad i \in \mathscr{Q}, \ t \in [s_2, s_1).$$

Therefore,

$$\dot{h}_{i_1}(t) \le \max_{i \in \mathscr{Q}} \{ M + \alpha_{i_1}, \ -a_{i_1 i} \frac{\beta}{\beta + h_i(t_2) - h_{i_1}(t_2)} \},$$

which is contradicted by $\lim_{t \to s_1} \dot{h}_{i_1}(t) = +\infty$. As a result, s_0 cannot be found and $h_j - h_i \le \beta$, $a_{ij} \ne 0$ holds for any $i, j \in \{1, 2, \ldots, N\}$.

Step 3: Proof of the stability of the closed-loop system.

Now, we have proved that (8.14) holds for $t \in [t_0, +\infty)$ due to the establishment of $h_j - h_i \le \beta$, $a_{ij} \ne 0$, $i, j = 1, 2, \ldots, N$. Thus, from the Lyapunov theory, we conclude the convergence of V_i and get $\lim_{t \to \infty} \|\Theta_i \hat{x}_i\| = 0$ and $\lim_{t \to \infty} \|\Theta_i e_i\| = 0$. Meanwhile, it holds that $\rho_i(u_i) \le M$, $i \in 1, 2, \ldots, N$.

Consider $h(t) = \max\{h_1(t), h_2(t), \ldots, h_N(t)\}$ which satisfies $\dot{h}(t) \le \max\{(M + \alpha_i)^{\frac{1}{h}} - \frac{1}{4\delta}, 0\}$. Then, one has $h_i(t) \le h(t) \le 4\delta(M + \alpha_i)$, $\lim_{t \to \infty} \|\hat{x}_i\| = 0$ and $\lim_{t \to \infty} \|e_i\| = 0$. It can be further seen that the asymptotic stability of system (8.1) under (8.9) is equivalent to that of (8.10). That is to say, $\lim_{t \to \infty} \|e_i\| = 0$ and $\lim_{t \to \infty} \|\hat{x}_i\| = 0$ if and only if $\lim_{t \to \infty} \|x_i\| = 0$.

Furthermore, it is shown that h_i is a bounded nondecreasing function, which implies that there exists a scalar h_i^* such that $\lim_{t \to \infty} h_i(t) = h_i^*$ for any $i = 1, 2, \ldots, N$. Meanwhile, it holds that $h_i(t) \le h_i^*$ for any t. Now, we prove that $h_i^* = h_j^*$ for any $i, j = 1, 2, \ldots, N$. We also prove this by contradiction. Suppose that there exists a pair (i, j) such that $h_i^* < h_j^*$. Since h_i is communicating through a strongly connected network $\mathscr{G}$, a path can be found from i to j, and there exists a pair (i_1, i_2) such that $h_{i_1}^* < h_{i_2}^*$, $a_{i_1 i_2} \ne 0$. It is noted that there exists a time instant T such that $h_{i_2}^* - \frac{1}{2}(h_{i_2}^* - h_{i_1}^*) \le h_{i_2}(t) \le h_{i_2}^*$ and $h_{i_1}^* - \frac{1}{2}(h_{i_2}^* - h_{i_1}^*) \le h_{i_1}(t) \le h_{i_1}^*$ for $t \ge T$. It can be obtained that

$$\dot{h}_{i_1} \ge -a_{i_1 i_2} \frac{h_{i_1}(t) - h_{i_2}(t)}{\beta + h_{i_1}(t) - h_{i_2}(t)}$$

$$\ge 3 a_{i_1 i_2} \frac{h_{i_2}^* - h_{i_1}^*}{2\beta + h_{i_1}^* - h_{i_2}^*}$$

$$> 0.$$

With this in mind, we have

$$h_{i_1}(t) \ge h_{i_1}^* - \frac{1}{2}(h_{i_2}^* - h_{i_1}^*) + 3 a_{i_1 i_2} \frac{h_{i_2}^* - h_{i_1}^*}{2\beta + h_{i_1}^* - h_{i_2}^*}(t - T),$$

and

$$\lim_{t \to \infty} h_{i_1}(t) = \infty,$$

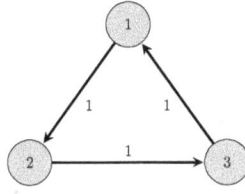

Figure 8.6   The communication between three controllers.

which is contradicted to the boundedness of $h_{i_1}(t)$. Thus, it holds that $h_i^* = h_j^*$, which means h_i and h_j converge to the same value, i, $j = 1, 2, \ldots, N$. This completes the proof.   □

8.5.3   SIMULATION

We consider system (8.5) and design the control law as

$$u_i = -0.3\frac{1}{h_i^3}x_{i,1} - 1.2\frac{1}{h_i^2}x_{i,2} - 0.7\frac{1}{h_i}x_{i,2}, \quad i = 1, 2, 3. \tag{8.15}$$

The communication network is shown as in Figure 8.6. Accordingly, the dynamic gains h_1, h_2, h_3 are designed as

$$\begin{aligned}
\dot{h}_1 &= \max\{\frac{1}{h_1}(u_1^2 + 2) - 0.25, -\frac{h_1 - h_2}{2 + h_1 - h_2}, 0\}, \quad h_1(0) = 1, \\
\dot{h}_2 &= \max\{\frac{1}{h_2}(u_2^2 + 2) - 0.25, -\frac{h_2 - h_3}{2 + h_2 - h_3}, 0\}, \quad h_2(0) = 1, \\
\dot{h}_3 &= \max\{\frac{1}{h_3}(u_3^2 + 2) - 0.25, -\frac{h_3 - h_1}{2 + h_3 - h_1}, 0\}, \quad h_3(0) = 1.
\end{aligned} \tag{8.16}$$

Then, the simulation results are depicted in Figure 8.7. It can be observed that the system states all converge to zero and the control gains converge to a consensus value, which is consistent with our theoretical analyses.

8.6   NOTES

This chapter has introduced variable gain control methods to stabilize a class of large-scale interconnected nonlinear systems. We have compared three different control methods, i.e., decentralized control, centralized control and distributed control. It is shown that distributed control has combined the advantages of both distributed control and decentralized control, which can ensure global system stability with a reduced communication cost. Simulation results have proven the effectiveness of the proposed methods.

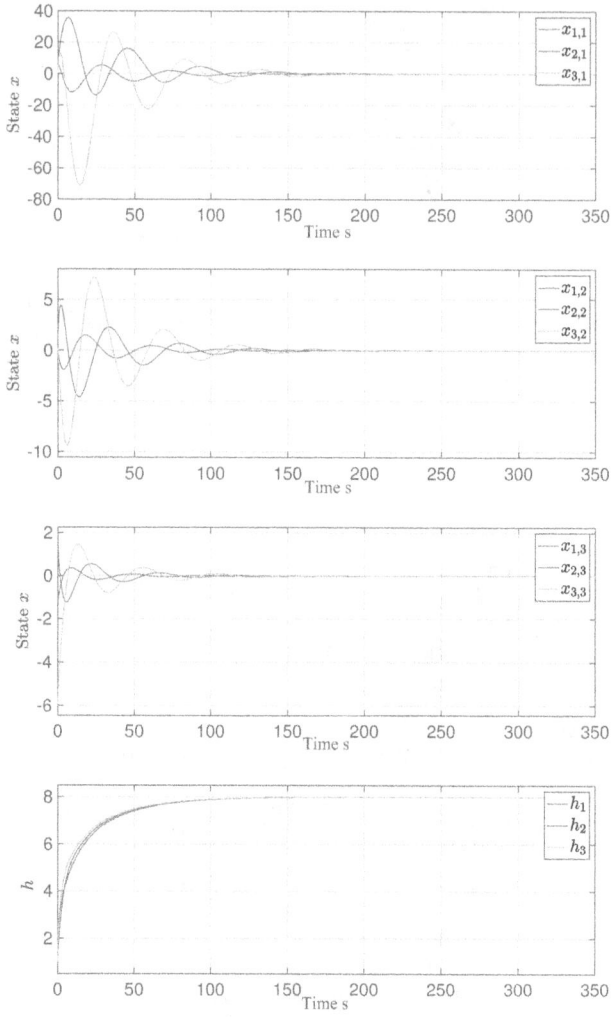

Figure 8.7   Trajectory of system (8.5), (8.15), (8.16).

9 Variable Gain Control for Feedforward Nonlinear Multi-Agent Systems

This chapter deals with the problem of control design for nonlinear multi-agent systems. The mathematical model of each agent has an upper triangular structure, with the nonlinear functions satisfying a Lipschitz growth condition. By introducing a suitable state transformation, the leader-follower consensus problem of the systems under consideration is converted into the control gain design problem. In particular, the gains are properly designed based on the Lyapunov analyses so that the proposed control method can achieve leader-follower consensus. An illustrative example is given to illustrate the effectiveness of the proposed control method.

9.1 BACKGROUND

In recent years, there has been an increasing research interest in the distributed control of multi-agent systems [109, 167], motivated by many applications in the field of biology, robotics, communications, and sensor networks. A critical problem for the distributed control of multi-agent systems is the consensus problem, which is to design appropriate control protocols so that agents can reach consensus by exchanging information over a shared communication network.

Many remarkable results have been reported for the consensus control problem of multi-agent systems [113, 125, 90, 41, 163, 68, 107]. Among these results, the consensus problem of nonlinear multi-agent systems is of particular interest because of its theoretical and practical importance. Due to the complexity of nonlinear systems, there is no particular control method that can be applied to all nonlinear multi-agent systems. Nevertheless, there are several effective control methods for nonlinear multi-agent systems with a certain structure, such as strict feedback and feedforward nonlinear multi-agent systems [127, 141, 144].

In this chapter, the leader-follower consensus control problem of feedforward nonlinear multi-agent systems is studied. The considered nonlinear dynamics satisfy a very general Lipschitz growth condition. When the Lipschitz constant is known, a distributed control method with a static gain is developed to achieve the leader-follower consensus; When the Lipschitz constant becomes unknown, a time-varying gain is developed that can still guarantee the leader-follower consensus. The effectiveness of the proposed control method is verified by a simulation example.

9.2   PROBLEM DESCRIPTION

NETWORK TOPOLOGY

The network topology of the N follower agents is by a simple graph $\mathscr{G}(\mathscr{V}, \mathscr{E}, \mathscr{A})$. The N agents are referred as the N nodes $\mathscr{V} = \{s_1, s_2, \ldots, s_N\}$. If agent i can receive information from agent j, we have $(s_i, s_j) \in \mathscr{E}$. Otherwise, $(s_i, s_j) \notin \mathscr{E}$. The graph is undirected, if once $e_{ij} \in \mathfrak{I}$, then $e_{ji} \in \mathfrak{I}$. The adjacency matrix is defined as $a_{ii} = 0$ and $a_{ij} = a_{ji} \geq 0 (i \neq j)$, where $a_{ij} > 0$ if and only if $e_{ij} \in \mathfrak{I}$. A path between s_i and s_j is a sequence of edges of the form $(s_k, s_{k+1}), k = i, i+1, \ldots, j-1$. The graph $\mathscr{G}$ is said to be connected if there exists a path between any two nodes of $\mathscr{G}$. The neighbor set of node s_i is defined as $N_i = \{s_j \in \mathscr{V} : e_{ij} \in \mathfrak{I}\}$. The degree of G is a diagonal matrix $\mathscr{D} = \mathrm{diag}(d_1, \ldots, d_N)$, where $d_i = \sum_{s_j \in N_i} a_{ij}$ for $i = 1, \ldots, N$. The Laplacian of graph $\mathscr{G}$ is denoted as $\mathscr{L} = \mathscr{D} - \mathscr{A}$. $\mathscr{H}$ is a subgraph of $\mathscr{G}$, if any two nodes of $\mathscr{H}$ are adjacent in $\mathscr{H}$ only if they are adjacent in $\mathscr{G}$. A subgraph $\mathscr{H}$ of $\mathscr{G}$ is called a component of $\mathscr{G}$ if it is maximal, subject to be connected.

We have another graph $\bar{\mathscr{G}}$ whose nodes set is $\{s_0\} \cup \mathscr{V}$, and s_0 represents the leader. The edges $e_{0j} = (s_0, s_j), j = 1, \ldots, N$, exist if and only if the agent j connects to the leader. The degree matrix of $\bar{\mathscr{G}}$ is denoted by $\mathscr{B} = \mathrm{diag}(b_1, \ldots, b_N)$, $b_i \geq 0$ is the adjacency weight between the agent i and the leader. If $b_i = 0$, it means that the agent i does not connect to the leader. $\bar{\mathscr{G}}$ is connected if at least one agent in each component is connected with the leader. Denote $\hat{\mathscr{L}} = \mathscr{L} + \mathscr{B}$. A useful lemma about $\hat{\mathscr{L}}$ is presented as follows:

Lemma 9.1

If graph $\bar{\mathscr{G}}$ is connected, then the symmetric matrix $\hat{\mathscr{L}}$ associated with $\bar{\mathscr{G}}$ is positive definite.   ∎

PROBLEM FORMULATION

For agent k, $k = 0, 1, \ldots, N$, the system model is described as

$$
\begin{cases}
\dot{x}_{k,1} = x_{k,2} + f_1(t, x_{k,3}, x_{k,4}, \ldots, x_{k,n}, u_k) \\
\dot{x}_{k,2} = x_{k,3} + f_2(t, x_{k,4}, \ldots, x_{k,n}, u_k) \\
\quad \vdots \\
\dot{x}_{k,n-2} = x_{k,n-1} + f_{n-2}(t, x_{k,n}, u_k) \\
\dot{x}_{k,n-1} = x_{k,n} + f_{n-1}(t, u_k) \\
\dot{x}_{k,n} = u_k
\end{cases}
\tag{9.1}
$$

where $x_{k,i} \in \mathbb{R}, i = 1, 2, \ldots, n$, $u_k \in \mathbb{R}$ represent the states, and input of the kth agent, respectively. $f_i, i = 1, 2, \ldots, n-1$, represents the nonlinear functions of the kth agent. Moreover, they satisfy the following Lipschitz assumption.

Assumption 9.1. *For $p,q = 0,1,2,\ldots,N$, any $(t,x_{p,i+2},\ldots,x_{p,n},u_p) \in \mathbb{R}^{n-i+1}$, and any $(t,x_{q,i+2},\ldots,x_{q,n},u_q) \in \mathbb{R}^{n-i+1}$, $i = 1,\ldots,n-1$, there exists a constant $\theta > 0$, such that*

$$|f_i(t,x_{p,i+2},\ldots,x_{p,n},u_p) - f_i(t,x_{q,i+2},\ldots,x_{q,n},u_q)|$$
$$\leq \theta \left(\sum_{j=i}^{n-1} |x_{p,j+2} - x_{q,j+2}| + |u_p - u_q| \right), \qquad (9.2)$$

where $x_{p,n+1} = x_{q,n+1} = 0$.

It is noted that system (9.1) can also be rewritten in the matrix form as

$$\dot{x}_k = Ax_k + Bu_k + f(t,x_k,u_k), \qquad (9.3)$$

where $x_k = (x_{k,1},x_{k,2},\ldots,x_{k,n})^{\mathrm{T}}$, and

$$A = \begin{pmatrix} 0 & 1 & 0 & \cdots & 0 \\ 0 & 0 & 1 & \cdots & 0 \\ \vdots & \vdots & \vdots & \ddots & \vdots \\ 0 & 0 & 0 & \cdots & 1 \\ 0 & 0 & 0 & \cdots & 0 \end{pmatrix}, \quad B = \begin{pmatrix} 0 \\ 0 \\ \vdots \\ 0 \\ 1 \end{pmatrix}, \quad f(t,x_k,u_k) = \begin{pmatrix} f_1(t,x_{k,3},x_{k,4},\ldots,x_{k,n},u_k) \\ f_1(t,x_{k,4},\ldots,x_{k,n},u_k) \\ \vdots \\ f_{n-1}(t,u_k) \\ 0 \end{pmatrix}.$$

The control objective is to propose a control method that can guarantee the leader-follower consensus of the system (9.1), i.e., $\lim_{t \to +\infty} \|x_k(t) - x_0(t)\| = 0$, $k = 1,\ldots,N$. To solve this problem, the following assumptions and lemma are needed.

Assumption 9.2. *All follower agents know the input of the leader.*

Assumption 9.3. *The communication topology of the $N+1$ agents is connected, i.e., the group $\mathscr{G}$ is connected.*

Lemma 9.2: [127, 125]

Suppose that $(\tilde{A},\tilde{B})$ is controllable, and $\lambda_1,\lambda_2,\ldots,\lambda_N$ are positive constants. Then, for any positive constant m, there exists a matrix K such that

$$\lambda_{\max}(\tilde{A} + \lambda_i \tilde{B}K) < -m, \quad i = 1,2,\ldots,N,$$

where $\lambda_{\max}(\tilde{A} + \lambda_i \tilde{B}K)$ represents the maximum eigenvalue of the matrix $\tilde{A} + \lambda_i \tilde{B}K$.

■

9.3 LOW-GAIN FEEDBACK CONTROL PROTOCOL

9.3.1 PROTOCOL DESIGN

We first consider the case where the Lipschitz constant θ is known. The control protocol is provided as follows:

$$u_k = KH \left(\sum_{j=1}^{N} a_{kj}(x_k - x_j) + b_k(x_k - x_0) \right) + u_0, \tag{9.4}$$

where $H = \text{diag}(\frac{1}{h^n}, \frac{1}{h^{n-1}}, \ldots, \frac{1}{h})$ with h being the static gain, and K is a vector to be determined.

For $k = 1, \ldots, N$, let $e_{k,i} = x_{k,i} - x_{0,i}$, $i = 1, \ldots, n$. Then, we obtain the error dynamics as

$$\dot{e}_{k,1} = e_{k,2} + \bar{f}_{k,1}$$
$$\dot{e}_{k,2} = e_{k,3} + \bar{f}_{k,2}$$
$$\vdots \tag{9.5}$$
$$\dot{e}_{k,n-1} = e_{k,n} + \bar{f}_{k,n-1}$$
$$\dot{e}_{k,n} = \Delta u_k$$

where $\Delta u_k = u_k - u_0$, and $\bar{f}_{k,i} = f_i(t, x_{k,i+2}, \ldots, x_{k,n}, u_k) - f_i(t, x_{0,i+2}, \ldots, x_{0,n}, u_0)$, $i = 1, \ldots, n-1$.

System (9.5) can also be written in the matrix form

$$\dot{e}_k = Ae_k + B\Delta u_k + \bar{f}_k \tag{9.6}$$

where $\bar{f}_k = (\bar{f}_{k,1}, \ldots, \bar{f}_{k,n-1}, 0)^T$, $e_k = (e_{k,1}, \ldots, e_{k,n})^T$, and A, B are given in (9.3).

Introducing a change of coordinate

$$\varepsilon_k = He_k, \quad k = 1, 2, \ldots, N$$

where $H = \text{diag}\{\frac{1}{h^n}, \frac{1}{h^{n-1}}, \ldots, \frac{1}{h}\}$, system (9.6) can be converted into

$$\dot{\varepsilon}_k = \frac{1}{h}A\varepsilon_k + \frac{1}{h}B\Delta u_k + F_k, \tag{9.7}$$

where $F_k = H\bar{f}_k$.

Denoting $\varepsilon = (\varepsilon_1^T, \ldots, \varepsilon_N^T)^T$, the above error dynamics can be further rewritten in the compact form

$$\dot{\varepsilon} = \frac{1}{h}I \otimes A\varepsilon + \frac{1}{h}\mathcal{L} \otimes BK\varepsilon + F, \tag{9.8}$$

where $F = (F_1^T, F_2^T, \ldots, F_N^T)^T$, and $\mathcal{L}$ is defined by the communication topology of the multi-agent system. Notice that under a connected network, $\mathcal{L}$ is positive definite. Thus, the eigenvalues of $\mathcal{L}$, i.e., $\lambda_1, \lambda_2, \ldots, \lambda_N$, are positive constants. From Lemma 9.2, we can find the vector K such that

$$\lambda_{\max}(A + \lambda_i BK) < 0, \quad i = 1, 2, \ldots, N. \tag{9.9}$$

9.3.2   CONSENSUS ANALYSIS

Theorem 9.1

Suppose that Assumptions 9.1–9.3 are satisfied. Let the vector K be provided by (9.9) and the controller be designed as (9.4), then the leader-follower consensus control problem of system (9.1) can be solved if $h \geq \max\{4\|P\|\beta, 1\}$.   ■

Proof. Under the consensus protocol (9.4), the closed-loop system is represented as (9.8). Since (9.9) holds, we can find a positive definite matrix P such that

$$P\left(I \otimes A\varepsilon + \mathscr{L} \otimes BK\varepsilon\right) + \left(I \otimes A\varepsilon + \mathscr{L} \otimes BK\varepsilon\right)^T P \leq -I.$$

We choose $V = \varepsilon^T P\varepsilon$, and the derivative of V along system (9.8) is given as

$$\dot{V}|_{(9.8)} = \frac{1}{h}\varepsilon^T\left((I_N \otimes A + \mathscr{L} \otimes BK)^T P + P(I_N \otimes A + \mathscr{L} \otimes BK)\right)\varepsilon + 2\varepsilon^T PF$$

$$\leq -\frac{1}{h}\|\varepsilon\|^2 + 2\varepsilon^T PF. \tag{9.10}$$

Noticing the definition of Δu_k in (9.4), we can get

$$|\Delta u_k|^2 = \left|KH\left(\sum_{j=1}^{N} a_{kj}(x_k - x_j) + b_k(x_k - x_0)\right)\right|^2$$

$$= \left|(\mathscr{A}_k\varepsilon)^T K^T K(\mathscr{A}_k\varepsilon)\right|$$

$$\leq \|K^T K\|\|\mathscr{A}_k^T \mathscr{A}_k\|\|\varepsilon\|^2,$$

where $\mathscr{A}_k$ is defined as $\mathscr{A}_k = \alpha_k \otimes I_n$, with α_k being the kth row of the matrix $\mathscr{L}$. Using (9.2) in Assumption 9.1 and $h \geq 1$, we have

$$\left|\frac{1}{h^{n+1-i}}\bar{f}_{k,i}\right| \leq \theta\frac{1}{h^{n+1-i}}\left(\sum_{j=i}^{n-1}|e_{k,j+2}| + |\Delta u_k|\right)$$

$$\leq \theta\frac{1}{h^2}\left(\sum_{j=i}^{n-1}|\varepsilon_{k,j+2}| + \sqrt{\|K^T K\|\|A_k^T A_k\|}\|\varepsilon\|\right)$$

$$\leq \frac{\theta}{h^2}\left(1 + \sqrt{\|K^T K\|\|A_k^T A_k\|}\right)\|\varepsilon\|$$

where $e_{k,n+1} = \varepsilon_{k,n+1} = 0$.

With the help of (9.7), from the definition of F in (9.8), we can find a constant β meeting

$$
\begin{aligned}
\|F\|^2 &= \sum_{k=1}^{N} F_k^T F_k \\
&= \sum_{k=1}^{N} (H \bar{f}_k)^T H \bar{f}_k \\
&= \sum_{k=1}^{N} (\|\bar{f}_{k,1}\|^2 + L^2 \|\bar{f}_{k,2}\|^2 + \cdots + L^{2n-4} \|\bar{f}_{k,n-1}\|^2) \\
&\le \frac{\beta^2}{h^4} \|\varepsilon\|^2.
\end{aligned}
$$

Back to (9.10), we have

$$
\dot{V}|_{(9.8)} \le -\frac{1}{h} \|\varepsilon\|^2 + 2\|P\| \frac{\beta}{h^2} \|\varepsilon\|^2.
$$

Thus, by choosing $h \ge \max\{4\|P\|\beta, 1\}$, we get $\dot{V}|_{(9.8)} \le -\frac{1}{2h}\|\varepsilon\|^2$, which means the exponential convergence of ε. Since h is a constant, it is established that the state $e_k(t)$ exponentially converges to zero, i.e., consensus errors e_k converge to zero. This ends the proof. $\qquad\square$

9.4 TIME-VARYING CONTROL PROTOCOL

9.4.1 PROTOCOL DESIGN

When the Lipschitz constant θ is unknown, the above method becomes invalid. To solve this problem, we propose a time-varying gain design method to construct the distributed controller. The control protocol is provided as follows:

$$
u_k = KH \left(\sum_{j=1}^{N} a_{kj}(x_k - x_j) + b_k(x_k - x_0) \right) + u_0, \tag{9.11}
$$

where $H = \text{diag}\{\frac{1}{h^n}, \frac{1}{h^{n-1}}, \ldots, \frac{1}{h}\}$ with $h = ct + 1$ being the time-varying gain. Both the vector K and the constant c will be designed later.

We also consider the error state $e_{k,i} = x_{k,i} - x_{0,i}$, $i = 1, \ldots, n$, $k = 1, 2, \ldots, N$. Then, we have

$$
\begin{aligned}
\dot{e}_{k,1} &= e_{k,2} + \bar{f}_{k,1} \\
\dot{e}_{k,2} &= e_{k,3} + \bar{f}_{k,2} \\
&\;\;\vdots \\
\dot{e}_{k,n-1} &= e_{k,n} + \bar{f}_{k,n-1} \\
\dot{e}_{k,n} &= \Delta u_k
\end{aligned} \tag{9.12}
$$

where $\Delta u_k = u_k - u_0$, and $\bar{f}_{k,i} = f_i(t, x_{k,i+2}, \ldots, x_{k,n}, u_k) - f_i(t, x_{0,i+2}, \ldots, x_{0,n}, u_0)$, $i = 1, \ldots, n-1$.

The consensus error dynamics (9.12) can be rewritten as

$$\dot{e}_k = Ae_k + B\Delta u_k + \bar{f}_k \tag{9.13}$$

where $\bar{f}_k = (\bar{f}_{k,1}, \ldots, \bar{f}_{k,n-1}, 0)^T$, $e_k = (e_{k,1}, \ldots, e_{k,n})^T$, and A, B are given in (9.3). Introducing a change of coordinate

$$\varepsilon_k = He_k, \quad k = 1, 2, \ldots, N,$$

system (9.13) can be converted into

$$\dot{\varepsilon}_k = \frac{1}{h}A\varepsilon_k + \frac{1}{h}B\Delta u_k - \frac{\dot{h}}{h}D\varepsilon_k + F_k,$$

where $F_k = H\bar{f}_k$, $D = \mathrm{diag}\{n, n-1, \ldots, 1\}$.

Denoting $\varepsilon = (\varepsilon_1^T, \ldots, \varepsilon_N^T)^T$, the above error dynamics can be further rewritten in the compact form

$$\dot{\varepsilon} = \frac{1}{h}I \otimes A\varepsilon + \frac{1}{h}\mathscr{L} \otimes BK\varepsilon - \frac{\dot{h}}{h}I \otimes D\varepsilon + F, \tag{9.14}$$

where $F = (F_1^T, F_2^T, \ldots, F_N^T)^T$. Since $\dot{h} = c$, we get

$$\dot{\varepsilon} = \frac{1}{h}I \otimes (A - cD)\varepsilon + \frac{1}{h}\mathscr{L} \otimes BK\varepsilon + F,$$

We can choose $c < \frac{1}{2n}$. Then, from Lemma 9.2, we can find the vector K such that

$$\lambda_{\max}(A - cD + \lambda_i BK) < -\frac{1}{2}, \quad i = 1, 2, \ldots, N. \tag{9.15}$$

9.4.2 CONSENSUS ANALYSIS

Theorem 9.2

Suppose that Assumptions 9.1–9.3 are satisfied. Let the vector K be provided by (9.15) and $c < \frac{1}{2n}$. Then, the leader-follower consensus control problem of the non-linear multi-agent system (9.1) can be solved by a time-varying consensus protocol (9.11).

∎

Proof. Let $V = \varepsilon^T P\varepsilon$ where P is the positive definite matrix satisfying

$$P(A - cD + \lambda_i BK + \frac{1}{2}I) + (A - cD + \lambda_i BK + \frac{1}{2}I)^T P \leq -I.$$

Then, we get

$$P(A - cD + \lambda_i BK) + (A - cD + \lambda_i BK)^T P \le -P,$$

The derivative of V along system (9.14) is given as

$$
\begin{aligned}
\dot{V}|_{(9.14)} =& \frac{1}{h}\varepsilon^T\left((I_N \otimes (A - cD) + \mathscr{L} \otimes BK)^T P + P(I_N \otimes (A - cD) + \mathscr{L} \otimes BK)\right)\varepsilon \\
& + 2\varepsilon^T PF \\
\le& -\frac{1}{h}\varepsilon P \varepsilon + 2\varepsilon^T PF.
\end{aligned}
$$

$$(9.16)$$

To estimate the term $2\varepsilon^T PF$, we can also get a constant β to satisfy

$$\|F\| \le \frac{\beta}{h^2}\|\varepsilon\|$$

where β is dependent on the constants $\|K^T K\|, \|A_k^T A_k\|$ and constant θ. Back to (9.16), we have

$$
\begin{aligned}
\dot{V}|_{(9.14)} \le& -\frac{1}{h}V + 2\|P\|\frac{\beta}{h^2}\|\varepsilon\|^2 \\
\le& -\frac{1}{h}V + 2\|P\|\frac{\beta}{h^2}\frac{1}{\lambda_{\min}(P)}V,
\end{aligned}
$$

where $\lambda_{\min}(P)$ is the minimum eigenvalue of P.

Let $\tilde{V} = h^{2n}V$, then we obtain

$$\dot{\tilde{V}}|_{(9.14)} \le -(1 - 2nc)\frac{1}{h}\tilde{V} + 2\|P\|\frac{\beta}{h^2}\frac{1}{\lambda_{\min}(P)}\tilde{V},$$

which further gives

$$
\begin{aligned}
\tilde{V}(t) \le& \left(\frac{1}{ct+1}\right)^{\frac{1-2nc}{c}} e^{2\beta\|P\|\frac{1}{c\lambda_{\min}(P)}\left(1 - \frac{1}{ct+1}\right)} V(0) \\
\le& \left(\frac{1}{ct+1}\right)^{\frac{1-2nc}{c}} e^{2\beta\|P\|\frac{1}{c\lambda_{\min}(P)}} \tilde{V}(0).
\end{aligned}
$$

Since

$$\lim_{t \to +\infty} \left(\frac{1}{ct+1}\right)^{\frac{1-2nc}{c}} = 0,$$

we have

$$\lim_{t \to +\infty} \tilde{V}(t) = 0.$$

It can be obtained that

$$\|e_k(t)\|^2 \le h^{2n}\|\varepsilon_k(t)\|^2 \le \frac{1}{\lambda_{\min}(P)}\tilde{V}(t).$$

Thus, $e_k(t)$ converges to zero, i.e., e_k converges to zero for all $k = 1, \ldots, N$. This ends the proof. $\qquad\square$

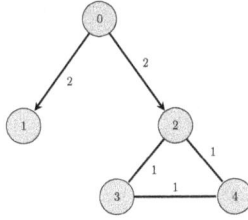

Figure 9.1 Communication topology.

9.4.3 SIMULATION

Consider a group of $4+1$ agents with identical nonlinear dynamics that are indexed by $0,1,2,3,4$. In this multi-agent system, the agent indexed by 0 is referred to as the leader and the agents indexed by $1,2,3,4$ are called the followers. For $k=0,1,2,3,4$, the agent k is described by

$$
\begin{aligned}
\dot{x}_{k,1} &= x_{k,2} + 10x_{k,3} + \frac{5}{2+e^t}\sin u_k \\
\dot{x}_{k,2} &= x_{k,3} + \frac{1}{2+\sin(t)}u_k \\
\dot{x}_{k,3} &= u_k.
\end{aligned}
\tag{9.17}
$$

The communication graph is shown in Figure 9.1. The adjacency matrix and degree matrix of the follower agents is denoted by $\mathscr{A}$ and $\mathscr{B}$, respectively. From Figure 9.1, $\mathscr{A}$ and $\mathscr{B}$ can be defined as follows:

$$
\mathscr{A} = \begin{pmatrix} 0 & 0 & 0 & 0 \\ 0 & 0 & 1 & 1 \\ 0 & 1 & 0 & 1 \\ 0 & 1 & 1 & 0 \end{pmatrix}, \quad
\mathscr{B} = \begin{pmatrix} 2 & 0 & 0 & 0 \\ 0 & 2 & 0 & 0 \\ 0 & 0 & 0 & 0 \\ 0 & 0 & 0 & 0 \end{pmatrix}.
$$

The consensus protocol is designed as

$$
u_k = KH\left(\sum_{j=1}^{3} a_{kj}(x_k - x_j) + b_k(x_k - x_0)\right) + u_0
\tag{9.18}
$$

where $K = (-6, -1, -3)$, $H = \mathrm{diag}(\frac{1}{h^3}, \frac{1}{h^2}, \frac{1}{h})$, $h = \frac{1}{7}t + 1$, a_{kj} is the element in the kth row and jth column of the matrix $\mathscr{A}$, and b_k is the kth diagonal element of the matrix $\mathscr{B}$. The initial conditions of these systems are given as

$$
\begin{aligned}
x_0(0) &= (-5.7, 4.2, -4.7)^T, & x_1(0) &= (0.4, -5.0, 3.0)^T, \\
x_2(0) &= (-2.8, -3.3, -2.9)^T, & x_3(0) &= (4.5, -3.5, -2.7)^T, \\
x_4(0) &= (-1.5, -5.0, -4.2)^T, & u_0 &= 5\sin(t) + 1.
\end{aligned}
$$

Figure 9.2 shows the state response of the closed-loop system consisting of (9.17) and (9.18). It can be seen that the states of each follower asymptotically tend to the states of the leader, which verifies the effectiveness of the proposed method. It can be seen in Figure 9.3 that the trajectory of input is also consensus.

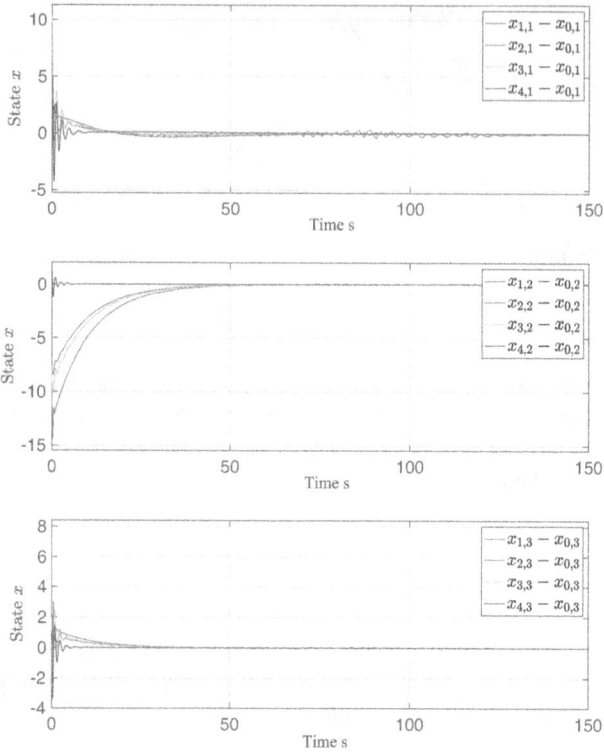

Figure 9.2   Consensus error state between the leader and followers.

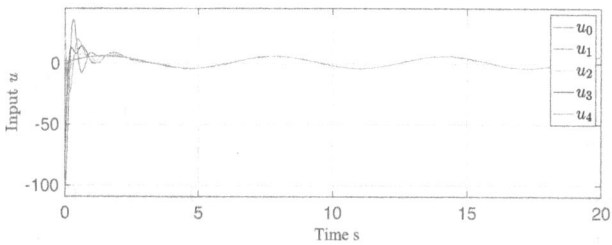

Figure 9.3   Inputs for the leader and followers.

9.5 NOTES

In this chapter, we have studied the leader-follower consensus control problem for a class of nonlinear multi-agent systems. We have developed two consensus control methods, namely the static gain method and the time-varying gain method. Based on the developed consensus methods, the states of all followers can track those of the leader. An example has been presented to demonstrate the effectiveness of the proposed methods.

10 Consensus Control for Nonlinear Multi-Agent Systems with Time Delays

In this chapter, a time-varying gain control approach is proposed for the leader-follower consensus issue of nonlinear multi-agent systems with distributed delays and discrete delays. It is assumed that each agent has the form of a strict feedback loop and satisfies the Lipschitz condition with an unknown constant. For each follower, a novel observer with time-varying gain is constructed. Using the observer, a distributed output feedback control protocol is designed for each follower agent under directed communication topology. It is proved that the consensus of multi-agent systems is achieved by the proposed control approach. The effectiveness of the proposed approach is tested by controlling two-stage chemical reactors.

10.1 BACKGROUND

Over the past decades, the consensus control of multi-agent systems has been widely studied due to its application in various fields, such as sensor networks and networked mobile robots [72] [118]. Leader-follower consensus is a fundamental problem in the collaborative control of nonlinear multi-agent systems. With the help of the backstepping technique, [10] and [43] designed different distributed control protocols which can ensure the states of each follower to track the states of the leaders. Based on the adaptive control method, the multi-agent systems under directed switching networks are considered in [83]. More recently, the gain construct approach has been applied to solving the leader-follower consensus problem of multi-agent systems, which avoids the complicated iterative procedure [150].

Time delays often limit system performance in engineering applications. Distributed delays are often used to model the time-delay phenomenon in some real control systems, such as propellant rocket motors and thermodynamics [157]. Many remarkable results have been obtained for systems with distributed delays [96]. Systems with discrete delays are also a hot research topic [62]. Notably, many works on linear multi-agent systems with time delays have been reported, see [81] and [164]. On the other hand, similar results for nonlinear multi-agent systems are limited [66] and [131].

In this chapter, a time-varying gain approach is proposed to solve the leader-follower consensus problem of nonlinear multi-agent systems with both distributed and discrete delays. The novelty of this chapter lies in the following two points:

- The first one is to propose a time-varying gain approach to design distributed output feedback controllers for nonlinear multi-agent systems, where each agent satisfies the Lipschitz condition with an unknown constant. Compared with the assumption imposed on the nonlinear functions in [66], the condition mentioned here is more general as the constant is unknown.

- The second one is to study nonlinear multi-agent systems with both distributed delays and discrete delays. It is difficult to choose a Lyapunov-Krasovskii function for handling distributed delays when the delay kernel is included in multi-agent systems. To deal with this difficulty, the Lyapunov-Razumikhin theorem is utilized, which facilitates the proof of system stability.

10.2 PROBLEM DESCRIPTION

NOTATION

In this chapter, the following notations are used.

- $U([-\tau, 0], \mathbb{R}^n)$ shows the set of continuous functions from $[-\tau, 0]$ to $\mathbb{R}^n$ for $\tau > 0$, and $U([-\tau, 0], \mathbb{R}^n) = \mathbb{R}^n$ for $\tau = 0$.
- For a continuous function $\zeta(t)$, let $\zeta^{(a)}(t) = \zeta(t - \tau)$ and $\zeta^{(b)}(t) = \int_0^\tau \phi(s)\zeta(t-s)ds$, where the delay kernel $\phi(s) : [0, \tau] \to \mathbb{R}$ is a measurable bounded function.
- $\|\zeta(t)\|$ is the Euclidean norm of the real vector function $\zeta(t)$, and $\|\zeta(t)\|_\tau = \sup_{-\tau \le \theta \le 0} \|\zeta(t+\theta)\|$.
- One introduces C to represent some known positive constants, and ϑ to express some unknown constants. Therefore, the expressions $C = C + C = C \cdot C$ and $\vartheta = \vartheta + \vartheta = \vartheta \cdot \vartheta = C + \vartheta = C \cdot \vartheta$ are correct. These constants are not the design parameters of the controllers, thus one does not need to consider the specific form of them.

GRAPH THEORY

Throughout this chapter, the following graph theory is presented ([66]).

A topology is formulated by $\mathscr{L} = (\mathscr{A}, \mathscr{B})$, where $\mathscr{A} = \{\mathscr{A}_1, \mathscr{A}_2, \dots, \mathscr{A}_N\}$ is the finite non-empty node set. $\mathscr{B} = \mathscr{A} \times \mathscr{A}$ means the topological edge set, and $(\mathscr{A}_i, \mathscr{A}_k) \in \mathscr{B}$ demonstrates that there exists an edge from nodes i to k. The adjacency or connectivity matrix is expressed as $\mathscr{C} = (\alpha_{i,k}) \in \mathbb{R}^{N \times N}$, where $\alpha_{i,k} > 0$ if $(\mathscr{A}_i, \mathscr{A}_k) \in \mathscr{B}$ and $\alpha_{i,k} = 0$ otherwise. The neighbors set of $\mathscr{A}_i$ is determined as $\mathscr{N}_i = \{\mathscr{A}_k | (\mathscr{A}_i, \mathscr{A}_k) \in \mathscr{B}\}$. The degree matrix of $\mathscr{L}$ is a diagonal matrix $\mathscr{D} = \text{diag}[v_1, v_2, \dots, v_N]$, where $v_i = \sum_{k \in \mathscr{N}_i} \alpha_{i,k}$. The Laplacian matrix is introduced as $\mathscr{L} = \mathscr{D} - \mathscr{C}$. The pinning matrix of $\mathscr{L}$ is denoted as $\mathscr{E} = \text{diag}[\beta_1, \beta_2, \dots, \beta_N]$, where $\beta_i > 0$ if there exists an edge from the leader to the ith follower and $\beta_i = 0$ otherwise. If a leader is indexed by 0, an augmented topology $\bar{\mathscr{L}}$ is acquired, which consists of the node 0, the graph $\mathscr{L}$ and edges between the leader and its neighbor agents. The graph matrix $\mathscr{H}$ is described as $\mathscr{H} = \mathscr{L} + \mathscr{E}$.

PROBLEM FORMULATION

We consider a group of $N+1$ agents consisting of one leader and N followers. The followers are referred as $1,2,\ldots,N$, and the leader is indexed by 0. The dynamics can be described as follows

$$
\begin{cases}
\dot{x}_{i,j} = x_{i,j+1} + f_j(t,\bar{x}_{i,j},\bar{x}_{i,j}^{(a)},\bar{x}_{i,j}^{(b)}), \ j=1,2,\ldots,n-1, \\
\dot{x}_{i,n} = u_i + f_n(t,\bar{x}_{i,n},\bar{x}_{i,n}^{(a)},\bar{x}_{i,n}^{(b)}), \\
y_i = x_{i,1}, \ i=0,1,\ldots,N,
\end{cases}
\tag{10.1}
$$

where $x_{i,k} \in \mathbb{R}$, $k=1,2,\ldots,n$, $u_i \in \mathbb{R}$ and $y_i \in \mathbb{R}$ represent the state, input and output of the ith agent, respectively; $u_0 = 0$; $\bar{x}_{i,k} = (x_{i,1},x_{i,2},\ldots,x_{i,k})^T$; $\bar{x}_{i,k}^{(a)} = (x_{i,1}^{(a)},x_{i,2}^{(a)},\ldots,x_{i,k}^{(a)})^T$ is the discrete-delay state; $\bar{x}_{i,k}^{(b)} = (x_{i,1}^{(b)},x_{i,2}^{(b)},\ldots,x_{i,k}^{(b)})^T$ is the distributed-delay state; the continuous nonlinear functions $f_k : \mathbb{R}^{1+3k} \to \mathbb{R}$, satisfy the following Lipschitz condition.

Assumption 10.1. For $k=1,2,\ldots,n$, there exists an unknown non-negative constant ϑ satisfying

$$
\begin{aligned}
& |f_k(t,\bar{x}_{i,k},\bar{x}_{i,k}^{(a)},\bar{x}_{i,k}^{(b)}) - f_k(t,\bar{x}_{0,k},\bar{x}_{0,k}^{(a)},\bar{x}_{0,k}^{(b)})| \\
& \leq \vartheta \sum_{m=1}^{k} (|x_{i,m}-x_{0,m}| + |x_{i,m}^{(a)}-x_{0,m}^{(a)}| + |x_{i,m}^{(b)}-x_{0,m}^{(b)}|).
\end{aligned}
\tag{10.2}
$$

Assumption 10.2. The topology $\mathscr{L}$ is directed and fixed, and $\mathscr{L}$ contains a spanning tree rooted at the leader.

10.3 DESIGN OF DISTRIBUTED OUTPUT FEEDBACK CONTROLLERS

In this section, the distributed output feedback controllers are designed. An observer for the ith follower is first constructed as:

$$
\begin{cases}
\dot{z}_{i,j} = z_{i,j+1} + c_j\gamma^j(\varepsilon_i - z_{i,1}), \ 1 \leq j \leq n-1, \\
\dot{z}_{i,n} = u_i + c_n\gamma^n(\varepsilon_i - z_{i,1}),
\end{cases}
\tag{10.3}
$$

where c_k, $k=1,2,\ldots,n$, are coefficients of the Hurwitz polynomial $\ell_1(\rho) = \rho^n + c_1\rho^{n-1} + \ldots + c_{n-1}\rho + c_n$, and the time-varying gain γ is given by

$$
\gamma = t + 2\tau + 1,
\tag{10.4}
$$

and ε_i is the output consensus error satisfying

$$
\varepsilon_i = \sum_{k \in \mathcal{N}_i} \alpha_{i,k}(y_i - y_k) + \beta_i(y_i - y_0),
\tag{10.5}
$$

where $\alpha_{i,k}$ and β_i are given in the graph theory.

Let

$$\rho_{i,k} = \frac{z_{i,k}}{\gamma^{\sigma+k-1}}, \quad k = 1, 2, \ldots, n, \tag{10.6}$$

with σ being a positive constant to be determined later.

It follows from (10.3) and (10.6) that

$$
\begin{cases}
\dot{\rho}_{i,j} = \gamma \rho_{i,j+1} + \dfrac{c_j}{\gamma^{\sigma-1}} \varepsilon_i - c_j \gamma \rho_{i,1} - (\sigma + j - 1)\dfrac{\dot{\gamma}}{\gamma}\rho_{i,j}, \quad j = 1, 2, \ldots, n-1, \\[2mm]
\dot{\rho}_{i,n} = \dfrac{u_i}{\gamma^{\sigma+n-1}} + \dfrac{c_n}{\gamma^{\sigma-1}} \varepsilon_i - c_n \gamma \rho_{i,1} - (\sigma + n - 1)\dfrac{\dot{\gamma}}{\gamma}\rho_{i,n}.
\end{cases}
\tag{10.7}
$$

The distributed output feedback controller for the ith follower is designed as:

$$u_i = -\gamma^{\sigma+n} \sum_{k=1}^{n} \delta_k \rho_{i,k}, \tag{10.8}$$

where δ_k, $k = 1, 2, \ldots, n$, are coefficients of the Hurwitz polynomial $\ell_2(\rho) = \rho^n + \delta_n \rho^{n-1} + \ldots + \delta_2 \rho + \delta_1$.

Based on (10.7) and (10.8), one obtains that

$$\dot{\rho}_i = \gamma \Upsilon \rho_i - \frac{\dot{\gamma}}{\gamma}\Delta \rho_i + \frac{1}{\gamma^{\sigma-1}} c \varepsilon_i, \tag{10.9}$$

where $\rho_i = (\rho_{i,1}, \rho_{i,2}, \ldots, \rho_{i,n})^T$, $c = (c_1, c_2, \ldots, c_n)^T$, $\Delta = \text{diag}(\sigma, \sigma + 1, \sigma + 2, \ldots, \sigma + n - 1)$, and

$$
\Upsilon = \begin{pmatrix}
-c_1 & 1 & \cdots & 0 \\
\vdots & \vdots & \ddots & \vdots \\
-c_{n-1} & 0 & \cdots & 1 \\
-c_n - \delta_1 & -\delta_2 & \cdots & -\delta_n
\end{pmatrix}.
$$

Define

$$\eta_{i,k} = x_{i,k} - x_{0,k} - z_{i,k},$$

and

$$\xi_{i,k} = \frac{\eta_{i,k}}{\gamma^{\sigma+k-1}}. \tag{10.10}$$

It can be seen from (10.1) and (10.3) that

$$\dot{\xi}_i = \gamma \Phi \xi_i - \frac{\dot{\gamma}}{\gamma}\Delta \xi_i - \frac{1}{\gamma^{\sigma-1}} c \varepsilon_i + \gamma c \rho_{i,1} + \Psi_i, \tag{10.11}$$

where
$\Psi_i = (\frac{1}{\gamma^\sigma}\tilde{f}_1, \frac{1}{\gamma^{\sigma+1}}\tilde{f}_2, \ldots, \frac{1}{\gamma^{\sigma+n-1}}\tilde{f}_n)^T$, $\tilde{f}_k = f_k(t, \bar{x}_{i,k}, \bar{x}_{i,k}^{(a)}, \bar{x}_{i,k}^{(b)}) - f_k(t, \bar{x}_{0,k}, \bar{x}_{0,k}^{(a)}, \bar{x}_{0,k}^{(b)})$,
$k = 1, 2, \ldots, n$, $\xi_i = (\xi_{i,1}, \xi_{i,2}, \ldots, \xi_{i,n})^T$, and

$$
\Phi = \begin{pmatrix}
0 & 1 & \cdots & 0 \\
\vdots & \vdots & \ddots & \vdots \\
0 & 0 & \cdots & 1 \\
0 & 0 & \cdots & 0
\end{pmatrix}.
$$

Let $\varepsilon_i = (\rho_i^T, \xi_i^T)^T$. It is obtained from (10.9) and (10.11) that

$$\dot{\varepsilon}_i = \gamma A \varepsilon_i - \frac{\dot{\gamma}}{\gamma} D \varepsilon_i - \frac{1}{\gamma^{\sigma-1}} \hat{c} \varepsilon_i + G_i, \tag{10.12}$$

where $\hat{c} = (-c^T, c^T)^T$, $G_i = (0_{1\times n}, \Psi_i^T)^T$, $\varsigma = (1, 0_{1\times(n-1)})^T$, and

$$A = \begin{pmatrix} \Upsilon & 0_{n\times n} \\ c\varsigma^T & \Phi \end{pmatrix}, \quad D = \begin{pmatrix} \Delta & 0_{n\times n} \\ 0_{n\times n} & \Delta \end{pmatrix}.$$

From (10.5), (10.6) and (10.10), one arrives at

$$\begin{aligned} \varepsilon_i &= \gamma^\sigma \sum_{k\in\mathcal{N}_i} \alpha_{i,k}(\xi_{i,1} + \rho_{i,1} - \xi_{k,1} - \rho_{k,1}) + \gamma^\sigma \beta_i(\xi_{i,1} + \rho_{i,1}) \\ &= \gamma^\sigma \Lambda^T \left(\sum_{k\in\mathcal{N}_i} \alpha_{i,k}(\varepsilon_i - \varepsilon_k) + \beta_i \varepsilon_i \right), \end{aligned} \tag{10.13}$$

where $\Lambda = (\varsigma^T, \varsigma^T)^T$.

Letting $\varepsilon = (\varepsilon_1^T, \varepsilon_2^T, \dots, \varepsilon_N^T)^T$, it follows from (10.12) and (10.13) that

$$\dot{\varepsilon} = \gamma B \varepsilon - \frac{\dot{\gamma}}{\gamma}(I_N \otimes D)\varepsilon + F, \tag{10.14}$$

where $B = I_N \otimes A - \mathcal{H} \otimes (\hat{c}\Lambda^T)$ and $F = (G_1^T, G_2^T, \dots, G_N^T)^T$.

Remark 10.1. *Since ε_i is introduced in the observer, the topology graph plays a significant role in achieving consensus. The observer (10.3) and the dynamic gain (10.4) only adopt the output information of the follower and its neighbors. Therefore, the distributed output feedback controller (10.8) is designed for each follower by only utilizing the output information of the follower and its neighbors. Compared with the work in [150], the controller (10.8) decreases the information transmission and economizes the communication bandwidth.*

10.4   CONSENSUS ANALYSIS

Based on the Lyapunov-Razumikhin theorem, the main results of this chapter are presented as follows.

Theorem 10.1

Under Assumptions 10.1-10.2 the leader-follower consensus of system (10.1) is achieved by the following distributed output feedback controllers

$$u_i = -\sum_{k=1}^{n} \delta_k \gamma^{n+1-k} z_{i,k}, \quad i = 1, 2, \dots, N,$$

where $z_{i,k}$, $k = 1, 2, \dots, n$, are given in (10.3), and $\gamma = t + 2\tau + 1$, and δ_k, $k = 1, 2, \dots, n$, are given in (10.8).   ∎

Proof. The matrix B is Hurwitz by choosing the suitable coefficients c_k and δ_k, $k = 1, 2, \ldots, n$, of the Hurwitz polynomials $\ell_1(\rho)$ and $\ell_2(\rho)$ ([66]). Therefore, there exists a positive definite matrix $P = \text{diag}(P_1, P_2, \ldots, P_N)$, $P_i \in \mathbb{R}^{2n \times 2n}$, $i = 1, 2, \ldots, N$, and a positive constant σ such that

$$B^T P + PB \le -I, \quad \sigma P_i \le DP_i + P_i D.$$

Setting a Lyapunov function candidate

$$V_\varepsilon = \varepsilon^T P \varepsilon,$$

one obtains

$$\dot{V}_\varepsilon|_{(10.14)} \le -\sum_{i=1}^{N}\left(\gamma\|\varepsilon_i\|^2 + \sigma\frac{\dot{\gamma}}{\gamma}\varepsilon_i^T P_i \varepsilon_i - 2\varepsilon_i^T P_i G_i\right). \tag{10.15}$$

By (10.2), (10.6) and (10.10), one has

$$\frac{1}{\gamma^{\sigma+k-1}}|\tilde{f}_k| \le \vartheta\sum_{m=1}^{k}\left(|\xi_{i,m}+\rho_{i,m}|+|\xi_{i,m}^{(a)}+\rho_{i,m}^{(a)}|+|\xi_{i,m}^{(b)}+\rho_{i,m}^{(b)}|\right).$$

Based on Young's inequality, we have

$$\|G_i\| \le \vartheta(\|\varepsilon_i\| + \|\varepsilon_i^{(a)}\| + \|\varepsilon_i^{(b)}\|). \tag{10.16}$$

Substituting (10.16) into (10.15) yields that

$$\dot{V}_\varepsilon|_{(10.14)} \le -\sum_{i=1}^{N}\left(\gamma\|\varepsilon_i\|^2 + \sigma\frac{\dot{\gamma}}{\gamma}\varepsilon_i^T P_i \varepsilon_i - \vartheta\|\varepsilon_i\|^2 - \|\varepsilon_i\|_\tau^2\right). \tag{10.17}$$

Taking a function $\ell_3(\rho) = q\rho$ with $q > 1$, one has $V_\varepsilon(\varepsilon(t+\theta)) < qV_\varepsilon(\varepsilon)$ and

$$\|\varepsilon\|_\tau^2 < \frac{q\lambda_{\max}(P)}{\lambda_{\min}(P)}\|\varepsilon\|^2.$$

It is inferred from (10.17) and $\gamma = t + 2\tau + 1$ that

$$\dot{V}_\varepsilon|_{(10.14)} \le -(t+2\tau+1)\sum_{i=1}^{N}\|\varepsilon_i\|^2 + \sum_{i=1}^{N}\frac{C}{t+2\tau+1}\|\varepsilon_i\|^2 + \vartheta\sum_{i=1}^{N}\|\varepsilon_i\|^2), \tag{10.18}$$
$$\text{if } V_\varepsilon(\varepsilon(t+\theta)) < qV_\varepsilon(\varepsilon).$$

There exists a finite time T_1 such that

$$t + 2\tau + 1 \ge \vartheta + C, \qquad t \ge T_1. \tag{10.19}$$

Substituting (10.19) into (10.18) yields

$$\dot{V}_\varepsilon|_{(10.14)} \le -CV_\varepsilon, \text{ for } t \ge T_2. \tag{10.20}$$

Solving the inequality (10.20), one has

$$V_\varepsilon \leq V_\varepsilon(T_2)e^{-Ct}.$$

Therefore, we obtain

$$
\begin{cases}
z_{i,k} = (t+2\tau+1)^{\sigma+k-1}\rho_{i,k} \leq \dfrac{C(t+2\tau+1)^C}{e^{Ct}}, \\[4mm]
\eta_{i,k} = (t+2\tau+1)^{\sigma+k-1}\xi_{i,k} \leq \dfrac{C(t+2\tau+1)^C}{e^{Ct}}.
\end{cases}
\tag{10.21}
$$

Based on L'Hopital's rule, one has

$$\lim_{t\to+\infty} |x_{i,k} - x_{0,k}| = 0, \quad k = 1,2,\ldots,n.$$

Furthermore, it follows from (10.8) and (10.21) that

$$
\begin{cases}
|u_i| \leq \dfrac{C(t+2\tau+1)^C}{e^{Ct}}, \\[4mm]
\lim_{t\to+\infty} u_i = \lim_{t\to+\infty} \dfrac{C(t+2\tau+1)^C}{e^{Ct}} = 0, \ i = 1,2,\ldots,N.
\end{cases}
$$

Therefore, one obtains that u_i, $i = 1,2,\ldots,N$, are bounded and converge to zero. The proof is completed. □

Remark 10.2. *The distributed delays and the discrete delays are considered in the system states, and it is difficult to select a Lyapunov-Krasovskii functional with a simple form to handle the distributed delays. Inspired by [62], when system (10.1) only has discrete delays, one can choose a single integral Lyapunov-Krasovskii functional $V = V_\varepsilon + C\sum_{i=1}^{N}\int_{t-\tau}^{t} e^{-C(t-s)}\|\varepsilon_i\|^2 ds$ such that $\dot{V} \leq -CV$, for $t \geq T_1$, where C represents some different positive constants. In order to avoid the complex forms of the Lyapunov-Krasovskii function, the Lyapunov-Razumikhin theorem is utilized effectively to simplify the proof process.*

10.5 SIMULATION

To show the effectiveness of the proposed consensus protocol, the two-stage chemical reactors with delayed recycle streams are considered

$$
\begin{cases}
\dot{v}_{i,1} = -\dfrac{1}{\varphi_{i,1}}v_{i,1} - \mu_{i,1}v_{i,1} + \dfrac{1-l_{i,2}}{\psi_{i,1}}v_{i,2} + g_1(\cdot), \\[4mm]
\dot{v}_{i,2} = -\dfrac{1}{\varphi_{i,2}}v_{i,2} - \mu_{i,2}v_{i,2} + \dfrac{l_{i,1}}{\psi_{i,2}}v_{i,1}(t-\tau) + \dfrac{m_{i,2}}{\psi_{i,2}}u_i + g_2(\cdot), \\[4mm]
y_i = v_{i,1}, \ i = 0,1,2,3,4,
\end{cases}
\tag{10.22}
$$

where the physical meanings of parameters are provided in Table 10.1. Due to the chemical reaction of fuels, the current states $v_{i,j}$, $j = 1,2$, of system (10.22) can

Table 10.1

Physical meanings of parameters

Parameters	Physical meanings
$m_{i,2}$	feed rate
$v_{i,1}, v_{i,2}$	compositions
g_1, g_2	nonlinear functions
$l_{i,1}, l_{i,2}$	recycle flow rates
$\mu_{i,1}, \mu_{i,2}$	reactor residence times
$\psi_{i,1}, \psi_{i,2}$	reactor volumes
$\varphi_{i,1}, \varphi_{i,2}$	reactor residence times

be affected by the states $v_{i,j}$, $j = 1, 2$, of the previous period of time. Therefore, it is reasonable to consider system (10.22) with both distributed delays and discrete delays.

The communication topology graph is illustrated in Figure 10.1. For simplicity, we suppose that $u_0 = 0$, $g_1 = 2\vartheta \int_0^\tau \phi(s) v_{i,1}(t - s)ds$, and $g_2 = \vartheta \sin(v_{i,2}(t - \tau))$, where ϑ is an unknown constant. The other parameters are set as $\varphi_{i,1} = \varphi_{i,2} = 10$, $\mu_{i,1} = 0.2$, $\mu_{i,2} = l_{i,1} = 0.1$, $\vartheta = l_{i,2} = m_{i,2} = \psi_{i,1} = \psi_{i,2} = 0.5$, $\tau = 1$, and $\phi(s) = s^2$.

According to Theorem 10.1 the following distributed output feedback controller for the ith follower is designed

$$
\begin{cases}
u_i = -(t+3)^2 z_{i,1} - 3(t+3)z_{i,2}, \\
\dot{z}_{i,1} = z_{i,2} + 2(t+3)(\varepsilon_i - z_{i,1}), \\
\dot{z}_{i,2} = u_i + (t+3)^2(\varepsilon_i - z_{i,1}).
\end{cases}
\tag{10.23}
$$

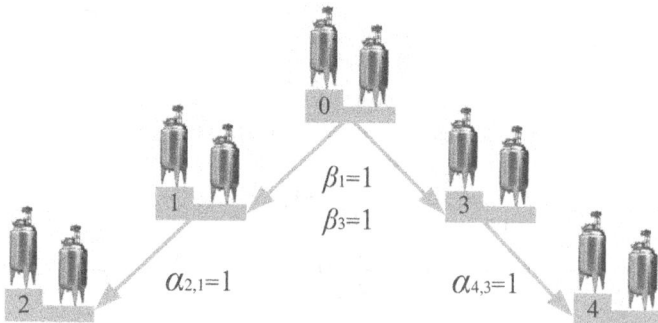

Figure 10.1   The communication topology.

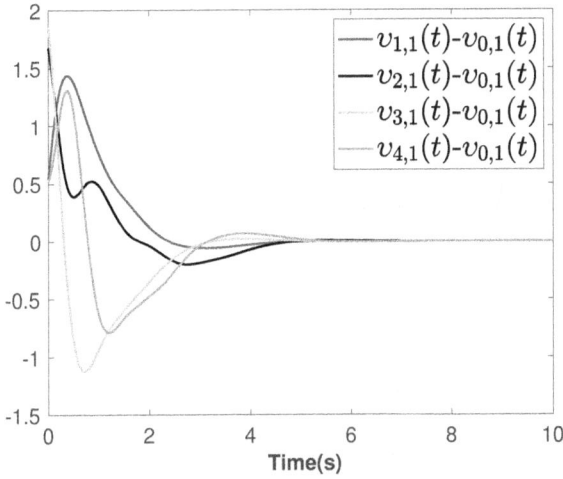

Figure 10.2   Trajectories of $v_{i,1}(t) - v_{0,1}(t)$.

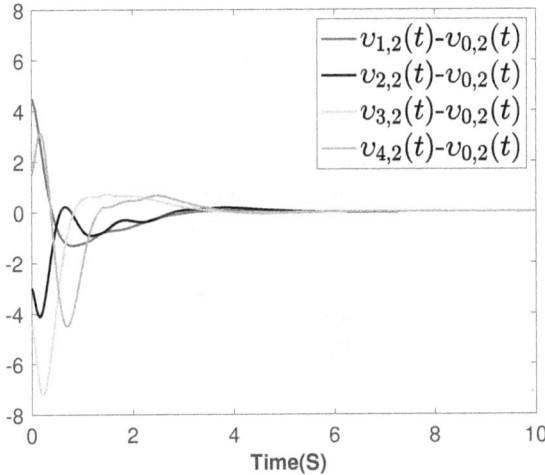

Figure 10.3   Trajectories of $v_{i,2}(t) - v_{0,2}(t)$.

Figures 10.2–10.4 show the control performance of the closed-loop system consisting of (10.22) and (10.23) with the initial condition $v_{0,1}(t) = -1$, $v_{0,2}(t) = 0.5$, $v_{1,1}(t) = -0.5$, $v_{1,2}(t) = 5$, $v_{2,1}(t) = 0.7$, $v_{2,2}(t) = -2.5$, $v_{3,1}(t) = 0.9$, $v_{3,2}(t) = -4$, $v_{4,1}(t) = -0.5$, $v_{4,2}(t) = 2$, and $u_i(t) = z_{i,1}(t) = z_{i,2}(t) = 0$, $\gamma_i(t) = 1$, $i = 1,2,3,4$, for $t \in [-1,0]$. From Figures 10.2 and 10.3, it is observed that all the

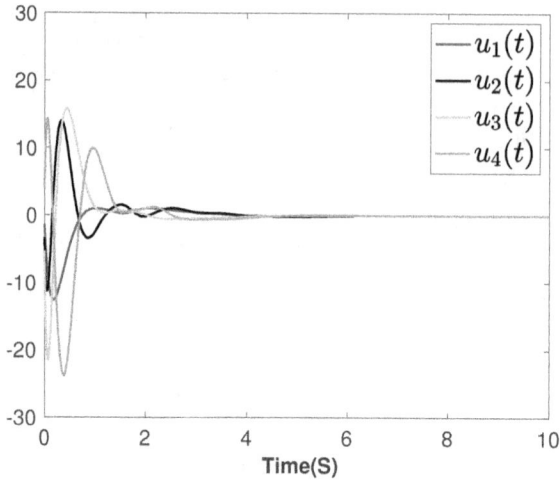

Figure 10.4 Trajectories of $u_i(t)$.

follower states can track the states of the leader, which demonstrates the effectiveness of the proposed control strategy.

10.6 NOTES

In this chapter, we have studied the leader-follower consensus problem via a time-varying gain approach for nonlinear multi-agent systems with both distributed and discrete delays. In contrast to existing approaches on nonlinear multi-agent systems, the proposed method for controller design does not involve the celebrated backstepping technique. The Lyapunov-Razumikhin theorem was used to deal with distributed delays and discrete delays, successfully avoiding the use of complex Lyapunov-Krasovskii functions. The effectiveness of the proposed method was demonstrated by controlling a group of two-stage chemical reactors.

Part IV

Application of Variable Gain
Control in Energy Conversion

11 Variable Gain Control of Three-Phase AC/DC Power Converters

For three-phase AC/DC converters, system uncertainties would cause large DC-link voltage fluctuations, further degrading the system performance or even leading to system instability. To deal with this problem, two kinds of controllers for DC-link voltage regulation are proposed in this chapter, namely, the static gain controller and the dynamic gain controller. The mathematical model of the AC/DC converter is built in the $\alpha\beta$ frame, where the control signal is able to decouple the regulation of active and reactive powers. Through a proper state transformation, the controller design is reduced to the design of the control gains. Subsequently, the stability of the AC/DC converter is theoretically proven. Finally, simulation results prove the effectiveness of the proposed static and dynamic gain control strategies.

11.1 BACKGROUND

With the increasingly tense conflicts in the areas of energy, economy, and environment, exploiting renewable energy has become a crucial task for all countries in the world [97, 116, 123]. Power electronic converters play an essential role in the development of renewable energy, as they serve as a channel for the conversion of renewable energy into electrical energy, which is illustrated in Figure 11.1. Due to the advantages of bidirectional power flow, high-quality grid-connected current and the adjustable power factor, three-phase AC/DC converters have been widely used in distributed power generation, energy storage systems, motor drive systems, etc. In order for AC/DC converters to generate the desired sinusoidal currents and provide stable DC-link voltages to the loads, the control strategy for AC/DC converters must be carefully designed [7, 34].

Many remarkable control approaches have been developed for AC/DC converters in recent decades. In most of these results, the AC/DC converters are controlled by a dual-loop control structure, i.e., the outer-voltage loop and the inner-current loop. In these two loops, traditional proportional-integral (PI) control is widely used because of its simplicity. The voltage controller regulates the DC-link voltage to its reference value and generates a reference signal for the inner-current loop. The inner-current loop regulates the active and reactive current and generates the PWM (pulse-width-modulation) signals to control the AC/DC converter. In essence, the AC/DC converter is a typical nonlinear system with the strong coupling of the system state. However, the PI controllers mentioned above are designed based on the linearized small-signal model around a predefined operating point. Therefore, the global

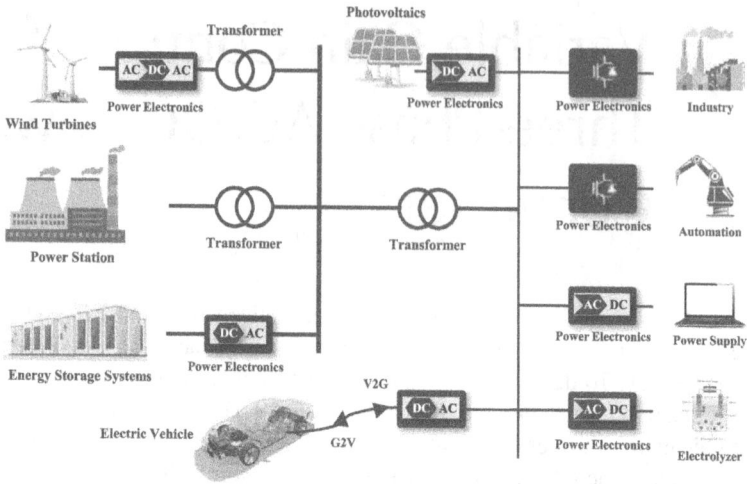

Figure 11.1   Power electronics in energy conversion.

large-signal stability cannot be rigorously guaranteed, resulting in a drop in performance when system uncertainties exist on AC and DC sides.

To solve the above problem, many nonlinear control methods for AC/DC converters based on large-signal nonlinear models have been proposed. Typical examples include adaptive control [139], sliding mode control (SMC) [79], backstepping control [124], and passivity-based control [24, 39]. Among these methods, SMC has received much attention due to its insensitivity to system uncertainties. However, the chattering phenomenon caused by SMC is difficult to handle, and can lead to a higher power and shorten the life of the switching elements. The super-twisting and high-order sliding mode control have been used to reduce the chattering phenomenon, but the chattering effects cannot be thoroughly eliminated. The backstepping control ensures good steady-state control performance. However, the stability analysis of backstepping control is performed based on an accurate mathematical model, making it invalid if system uncertainties exist. Moreover, the backstepping method has a complicated iterative process of control design. Based on these observations, it is reasonable to develop a simple and robust control for AC/DC converters.

This chapter proposes two control schemes for AC/DC converters, namely, static gain control and dynamic gain control. The mathematical model of the AC/DC converter is constructed in the $\alpha\beta$ frame so that active and reactive power are regulated directly. By introducing a suitable state transformation, the static and dynamic gain control strategies reduce the control design problem to the problem of determining the control gains, thus avoiding the iterative process of control. Meanwhile, these two methods have a single-loop control structure, which is simpler than the traditional dual-loop structure. Next, the stability of the AC/DC converter is theoretically demonstrated. Unlike static gain control, dynamic gain control designs a dynamic

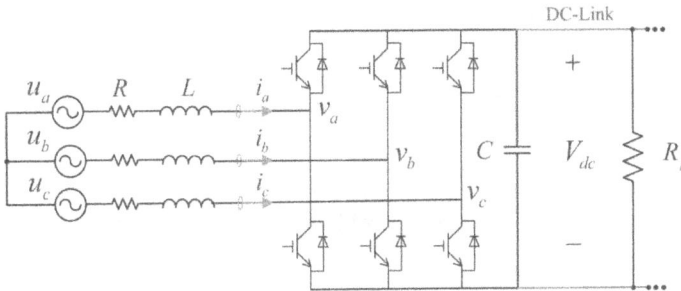

Figure 11.2 The structure of the three-phase AC/DC converter.

gain law instead of a constant gain, which ensures fast and accurate control under variable working conditions and load disturbances.

11.2 MATHEMATICAL MODEL AND PRELIMINARIES

Figure 11.2 shows the circuit structure of a three-phase AC/DC converter connected to the grid with L-filters. According to the law of conservation of power and ignoring the power losses, the DC-link power is described as

$$P = V_{dc}\frac{V_{dc}}{R_L} + CV_{dc}\dot{V}_{dc} \tag{11.1}$$

where P is the active power of the converter and V_{dc} is the DC-link voltage. C and R_L represent DC-link capacitor and load resistance, respectively. Using (11.1), the DC-link model is given as

$$\dot{V}_{dc} = \frac{P}{CV_{dc}} - \frac{V_{dc}}{CR_L}. \tag{11.2}$$

In the $\alpha\beta$ frame, the instantaneous active power P and reactive power Q on the AC-side are defined as

$$P = \frac{3}{2}u_\alpha i_\alpha + \frac{3}{2}u_\beta i_\beta$$

$$Q = \frac{3}{2}u_\beta i_\alpha - \frac{3}{2}u_\alpha i_\beta$$

where $u_{\alpha,\beta}$ and $i_{\alpha,\beta}$ are the grid voltages and currents in the $\alpha\beta$ frame, respectively.

In order to directly regulate the powers of the AC/DC converter without any phase-locked loop (PLL) involved, the AC-side model in the $\alpha\beta$ frame is given as [42]

$$\dot{P} = -\frac{3}{2L}u_\alpha v_\alpha - \frac{3}{2L}u_\beta v_\beta - \omega Q - \frac{R}{L}P + \frac{3}{2L}\left(u_\alpha^2 + u_\beta^2\right)$$

$$\dot{Q} = \frac{3}{2L}u_\alpha v_\beta - \frac{3}{2L}u_\beta v_\alpha + \omega P - \frac{R}{L}Q \tag{11.3}$$

where v_α and v_β are the converter voltages in the $\alpha\beta$ frame. ω is the grid angle frequency and R is the equivalent resistance of the power transmission line.

Based on (11.3), an improved ac-side model is built as

$$\dot{P} = -\frac{3}{2L}u_1 - \omega Q - \frac{R}{L}P + \frac{3}{2L}\left(u_\alpha^2 + u_\beta^2\right)$$
$$\dot{Q} = \frac{3}{2L}u_2 + \omega P - \frac{R}{L}Q \tag{11.4}$$

where the introduced new system control inputs are defined as

$$\begin{pmatrix} u_1 \\ u_2 \end{pmatrix} = \begin{pmatrix} u_\alpha & u_\beta \\ -u_\beta & u_\alpha \end{pmatrix} \begin{pmatrix} v_\alpha \\ v_\beta \end{pmatrix}.$$

Compared with [42], the control inputs u_1 and u_2 are introduced in (11.4) to decouple the original control inputs v_α and v_β. Subsequently, it is able to regulate the active and reactive powers separately [33, 26].

Combining (11.2) and (11.4), the model of the AC/DC converter is given as

$$\dot{V}_{dc} = \frac{P}{CV_{dc}} - \frac{V_{dc}}{CR_L}$$
$$\dot{P} = -\frac{3}{2L}u_1 - \omega Q - \frac{R}{L}P + \frac{3}{2L}\left(u_\alpha^2 + u_\beta^2\right).$$
$$\dot{Q} = \frac{3}{2L}u_2 + \omega P - \frac{R}{L}Q$$

For simplicity, the following state variables are defined as $x_1 = V_{dc}^2, x_2 = P, x_3 = Q$. Then, the dynamic model of the AC/DC converter can be rewritten as

$$\dot{x}_1 = \frac{2}{C}x_2 - \frac{2}{CR_L}x_1$$
$$\dot{x}_2 = -\frac{3}{2L}u_1 - \frac{R}{L}x_2 - \omega x_3 + \frac{3}{2L}\left(u_\alpha^2 + u_\beta^2\right). \tag{11.5}$$
$$\dot{x}_3 = \frac{3}{2L}u_2 - \frac{R}{L}x_3 + \omega x_2$$

In order to realize stable DC-link voltage and unit power factor, the control objectives are given as

- *I*: constructing u_1 such that the dc-link voltage V_{dc} tracks its reference V_{dc}^*;
- *II*: constructing u_2 such that Q tracks $Q^* = 0$.

11.3   DC-LINK VOLTAGE CONTROL

In this section, two different controllers are designed for DC-link voltage regulation, namely, a static gain controller and a dynamic gain controller.

11.3.1 STATIC GAIN CONTROL DESIGN

Firstly, the state transformation is introduced as

$$
\begin{aligned}
z_1 &= x_1 - (V_{dc}^*)^2 \\
z_2 &= \frac{1}{H_1}\left(x_2 - \frac{1}{R_L}(V_{dc}^*)^2\right)
\end{aligned}
\tag{11.6}
$$

where H_1 is a constant control gain and will be designed later.

Taking the derivative of $z = [z_1, z_2]^T$ yields

$$
\begin{aligned}
\dot{z}_1 &= \frac{2}{C}H_1 z_2 - \frac{2}{CR_L}z_1 \\
\dot{z}_2 &= \frac{1}{H_1}\left(-\frac{3}{2L}u_1 - \frac{R}{L}x_2 - \omega x_3 + \frac{3}{2L}\left(u_\alpha^2 + u_\beta^2\right)\right).
\end{aligned}
\tag{11.7}
$$

According to the static gain control method, the DC-link voltage controller is constructed as:

$$
u_1 = -\frac{2L}{3}\left(-H_1^2\left(k_1 z_1 + k_2 z_2\right) + \frac{R}{L}x_2 + \omega x_3 - \frac{3}{2L}\left(u_\alpha^2 + u_\beta^2\right)\right).
\tag{11.8}
$$

Substituting (11.8) into (11.7), one has

$$
\dot{z} = H_1 Az - G
$$

where

$$
A = \begin{pmatrix} 0 & \frac{2}{C} \\ -k_1 & -k_2 \end{pmatrix}, \qquad G = \begin{pmatrix} \frac{2}{CR_L}z_1 \\ 0 \end{pmatrix}.
$$

If k_1 and k_2 are selected to make A Hurwitz, then there exists a positive definite Lyapunov weighting matrix M such that

$$
MA + A^T M < -I
$$

where I is an identity matrix.

Choose the Lyapunov function as $V_1 = z^T M z$, then the derivative of V_1 is given by:

$$
\begin{aligned}
\dot{V}_1 &= \dot{z}^T M z + z^T M \dot{z} \\
&= (H_1 Az - G)^T M z + z^T M (H_1 Az - G) \\
&= H_1 z^T \left(A^T M + MA\right) z - 2G^T M z \\
&\leq -H_1 \|z\|^2 - 2z^T M G \\
&\leq -\left(H_1 - \frac{4}{CR_L}\|M\|\right)\|z\|^2.
\end{aligned}
$$

If $H_1 > \frac{4}{CR_L}\|M\| + n_0$, z_1 tends to zero according to the Lyapunov stability theory, i,e., objective I is achieved.

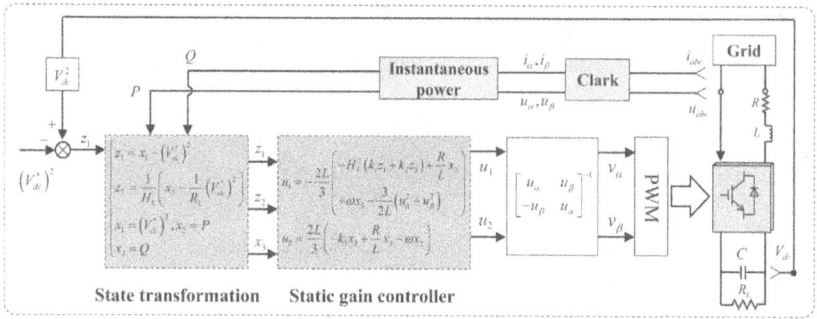

Figure 11.3   The control block diagram of the proposed static gain control method.

Take the Lyapunov function as

$$V_Q = \frac{1}{2}x_3^2.$$

Then the derivative of V_Q is given by

$$\dot{V}_Q = x_3 \left(\frac{3}{2L}u_2 - \frac{R}{L}x_3 + \omega x_2 \right). \tag{11.9}$$

Construct the reactive power controller as

$$u_2 = \frac{2L}{3} \left(-k_3 x_3 + \frac{R}{L}x_3 - \omega x_2 \right) \tag{11.10}$$

where k_3 is a positive constant.

Substituting (11.10) into (11.9), one has

$$\dot{V}_Q = -k_3 x_3^2.$$

Similarly, x_3 tends to zero when the time goes to infinity, implying that Objective II is achieved.

Figure 11.3 shows the control block diagram of the proposed static gain control method. The main conclusion of this subsection is summarized in the following theorem.

Theorem 11.1

For the AC/DC converter system (11.5), by using the static state transformation (11.6) and the controllers (11.8) and (11.10), the control objectives I-II are achieved.   ∎

11.3.2 DYNAMIC GAIN CONTROL DESIGN

This subsection constructs a dynamic gain controller for DC-link voltage regulation of the AC/DC converter. Firstly, define the following state variables as:

$$\chi_1 = x_1 - (V_{dc}^*)^2$$
$$\chi_2 = x_2 - \frac{1}{R_L}(V_{dc}^*)^2.$$

Then, a dynamic state transformation is introduced as:

$$\kappa_1 = \frac{1}{r}\chi_1$$
$$\kappa_2 = \frac{1}{r^2}\chi_2$$
$$\tag{11.11}$$

where r is a dynamic control gain and will be designed later.

Taking the derivative of (11.11) yields

$$\dot{\kappa}_1 = -\frac{2}{CR_L}\kappa_1 + r\frac{2}{C}\kappa_2 - \frac{\dot{r}}{r}\kappa_1$$
$$\dot{\kappa}_2 = \frac{1}{r^2}\left(-\frac{3}{2L}u_1 - \frac{R}{L}x_2 - \omega x_3 + \frac{3}{2L}\left(u_\alpha^2 + u_\beta^2\right)\right) - 2\frac{\dot{r}}{r}\kappa_2.$$

Using the dynamic gain control method, the dynamic gain controller is constructed as:

$$u_1 = -\frac{2L}{3}\left(-r^3(k_1\kappa_1 + k_2\kappa_2) + \frac{R}{L}x_2 + \omega x_3 - \frac{3}{2L}\left(u_\alpha^2 + u_\beta^2\right)\right). \tag{11.12}$$

Based on (11.12), the derivative of (11.11) is rewritten as:

$$\dot{\kappa}_1 = -\frac{2}{CR_L}\kappa_1 + r\frac{2}{C}\kappa_2 - \frac{\dot{r}}{r}\kappa_1$$
$$\dot{\kappa}_2 = -rk_1\kappa_1 - rk_2\kappa_2 - 2\frac{\dot{r}}{r}\kappa_2$$

and one has

$$\dot{\kappa} = rA\kappa + \phi - \frac{\dot{r}}{r}D\kappa$$

where $\kappa = (\kappa_1, \kappa_2)^T$, and

$$A = \begin{pmatrix} 0 & \frac{2}{C} \\ -k_1 & -k_2 \end{pmatrix}, D = \begin{pmatrix} 1 & 0 \\ 0 & 2 \end{pmatrix}, \phi = \begin{pmatrix} -\frac{2\kappa_1}{CR_L} \\ 0 \end{pmatrix}.$$

There exist a positive definite matrix N and constants k_1, k_2 such that

$$NA + A^T N < -I, \quad \alpha I \le ND + DN \le \beta I$$

where I is an identity matrix, $\alpha > 0$, and $\beta > 0$.

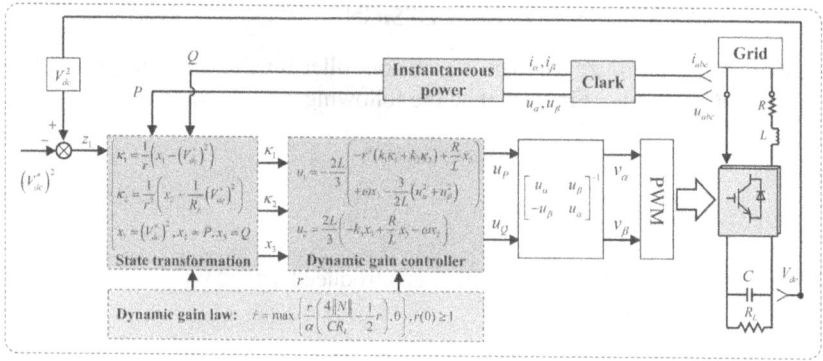

Figure 11.4 The control block diagram of the proposed dynamic gain control method.

Construct the dynamic gain law as

$$\dot{r} = \max\left\{\frac{r}{\alpha}\left(\frac{4\,\|N\|}{CR_L} - \frac{1}{2}r\right), 0\right\}, r(0) \geq 1. \tag{11.13}$$

It is obtained that $r(t) \geq r(0) \geq 1$. If $r > \frac{8\|N\|}{CR_L}$, $\dot{r} = 0$, so r is a bounded signal. Choose the Lyapunov function as

$$V_2 = \kappa^T N \kappa.$$

Then the derivative of V_2 is given by

$$\dot{V}_2 = \dot{\kappa}^T N \kappa + \kappa^T N \dot{\kappa}$$

$$= \left(rA\kappa + \phi - \frac{\dot{r}}{r}D\kappa\right)^T N\kappa + \kappa^T N\left(rA\kappa + \phi - \frac{\dot{r}}{r}D\kappa\right)$$

$$= r\kappa^T\left(A^T N + NA\right)\kappa + 2\kappa^T P\phi - \frac{\dot{r}}{r}\kappa^T\left(ND + DN\right)\kappa$$

$$\leq -r\|\kappa\|^2 + 2\|x\|\,\|N\|\,\|\phi\| - \alpha\frac{\dot{r}}{r}\|\kappa\|^2$$

$$\leq -r\|\kappa\|^2 + \frac{4\,\|N\|}{CR_L}\|\kappa\|^2 - \alpha\frac{\dot{r}}{r}\|\kappa\|^2$$

$$\leq -\frac{1}{2}r\|\kappa\|^2.$$

According to the Lyapunov stability theory, V_2 tends to zero when time goes to infinity.

Figure 11.4 shows the control block diagram of the proposed dynamic gain control method. The main conclusion in this subsection is summarized as follows:

Table 11.1

Parameters of the tested AC/DC converter.

Parameter	Symbol	Value
Rated power	P_0	2 kW
Filter inductance	L	1 mH
Line resistance	R	0.1 Ω
DC-link load resistance	R_L	60 Ω
DC-link voltage	V_{dc}	220 V
Grid voltage	$U_{a,b,c}$	100 V
Grid angle frequency	ω	100π rad/s
Switching frequency	f_{sw}	10 kHz
Sampling frequency	f_{sa}	10 kHz

Theorem 11.2

For the AC/DC converter system (11.5), by using the dynamic state transformation (11.11), the controllers (11.12) and (11.10), and the dynamic gain law (11.13), the control objectives I-II are achieved. ■

11.4 CASE STUDIES

In order to verify the effectiveness of the proposed two control methods, case studies are conducted in MATLAB/Simulink. It should be noted that k_1 and k_2 are chosen to be the same for these two methods. Table 11.1 gives the parameters of the AC/DC converter.

Figures 11.5–11.7 show the control performance when the AC/DC converter system changes from uncontrolled rectifying to controlled rectifying at $Q^* = 0$. To be specific, Figure 11.5 gives the curves of the DC-link voltage with a low gain $H_1 = 30$, a high gain $H_1 = 100$, and a dynamic gain determined by the dynamic gain control. Figure 11.6 and Figure 11.7 show the responses of grid currents and the dynamic gain r, respectively. It is observed that the static gain control with a high gain and dynamic gain control both obtain good control performances in the steady state. The DC-link voltage overshoot is 2.7 V and the settling time is 19 ms for the static gain control with a low gain, while the static gain control with the high-gain method achieves a faster dynamic performance (8 ms) but with a larger overshoot (4.3 V). Therefore, for the static gain control, the high gain brings a large DC-link voltage overshoot, while the low gain decreases the dynamic response. Besides, the high

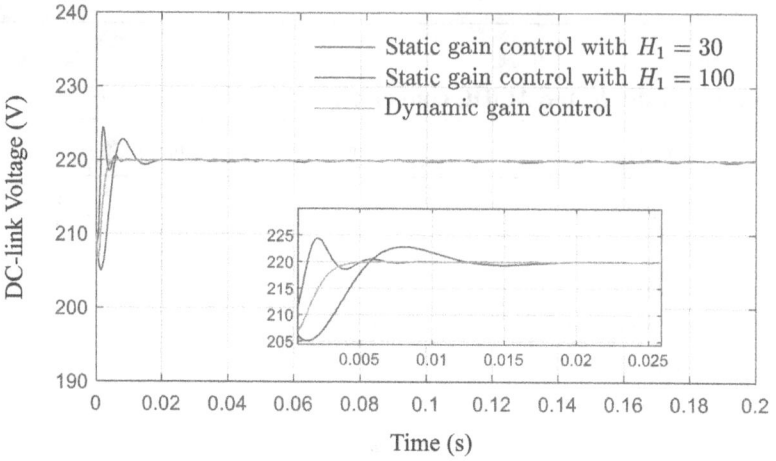

Figure 11.5   Curves of the DC-link voltage.

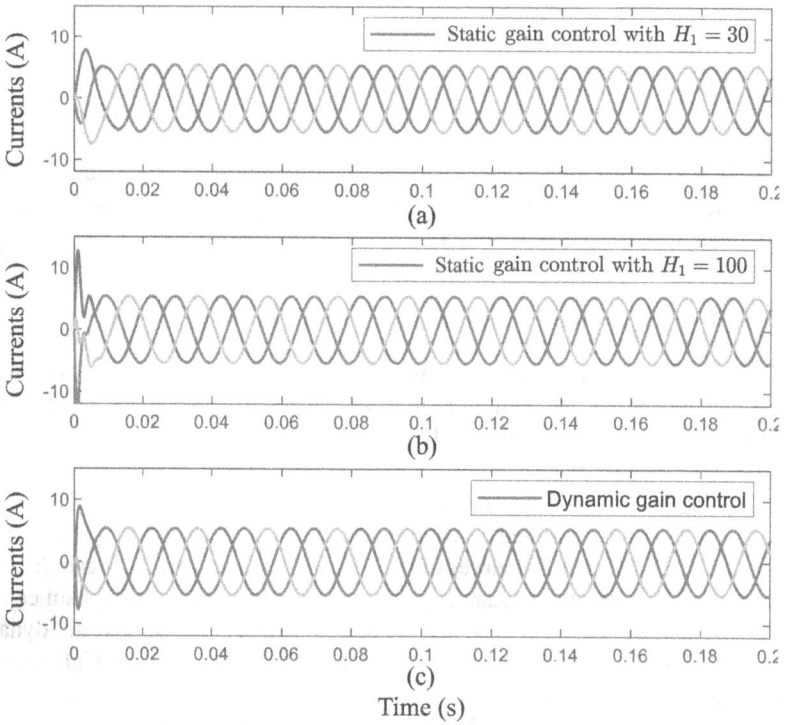

Figure 11.6   Curves of the grid currents.

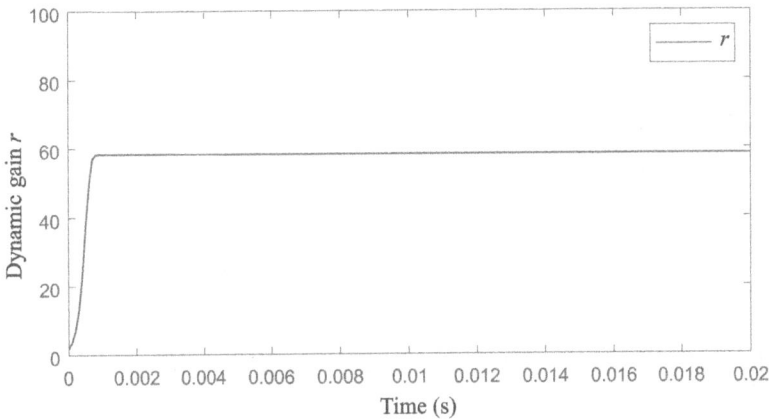

Figure 11.7 Response of the dynamic gain r.

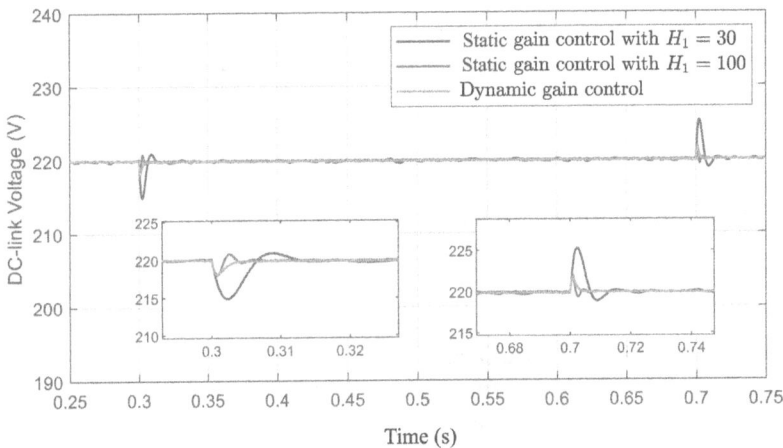

Figure 11.8 Responses of the DC-link voltage under load disturbances.

gain may cause large fluctuations of the currents, threatening the safety and stability of the AC/DC converter. The dynamic gain control achieves the fastest convergence speed (6 ms) without overshoot, which benefits from the design of the dynamic gain law.

In this test, the load R_L changes from 60 Ω to 40 Ω at $t = 0.3$ s and then recovers to 60 Ω at $t = 0.7$ s, and the transient response waveforms are captured in Figures 11.8–11.9. It can be seen that all control strategies can regulate the DC-link voltage to its reference value. However, compared with the static gain control with a low gain, the dynamic gain control and the static gain control with a high gain obtain a faster convergence speed and smaller fluctuations of the DC-link voltage.

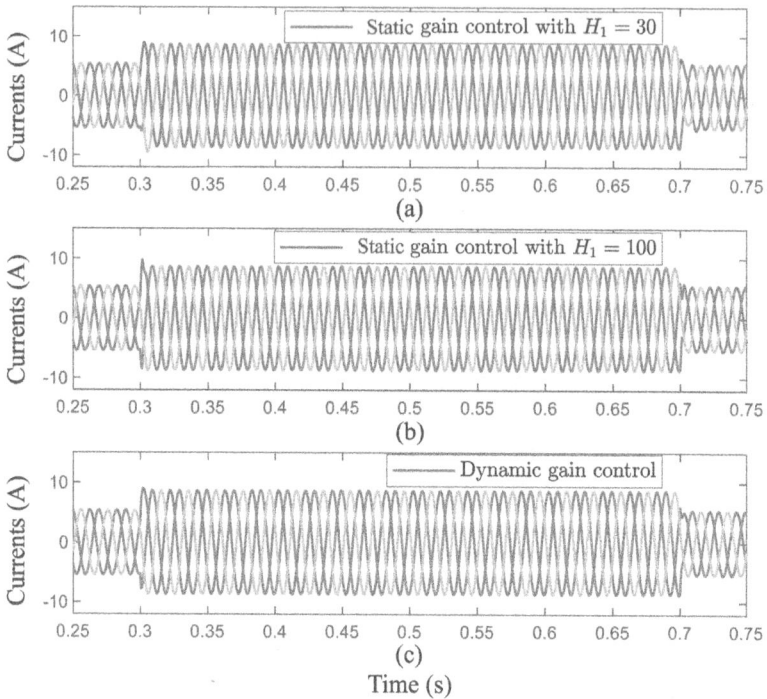

Figure 11.9   Responses of currents under load disturbances.

11.5   NOTES

In this chapter, static gain control and dynamic gain control strategies are successfully applied in the AC/DC converter system. After introducing a static/dynamic state transformation, the controller design problem becomes a gain design one. For these two methods, the control structures both include a state transformation and a single-loop controller, which is simpler than the traditional dual-loop control structure. In addition, the proposed methods are designed in the $\alpha\beta$ frame, which eliminates the need to use PLL and further simplifies the control design. Finally, the effectiveness of the proposed two methods is verified and compared under different case studies.

12 Variable Gain Control for Permanent Magnet Synchronous Motor Speed Regulation

This chapter proposes a fixed-time dynamic gain control to solve the speed regulation problem of a permanent magnet synchronous motor (PMSM). First, the original PMSM model is transformed into a new model by a dynamic state transformation. In this new model, the control coefficients are regulated by two auxiliary parameters. Then, a fixed-time controller is presented by properly designing the dynamics of the two auxiliary parameters. Rigorous closed-loop stability analysis is performed based on the Lyapunov theory. Finally, several case studies are conducted under different working conditions to verify the advantages of the proposed control strategy.

12.1 BACKGROUND

The permanent magnet synchronous motor (PMSM) is one of the most competitive motion control products due to its inherent advantages such as low rotor inertia, high power density, and compact size. A more detailed description of PMSMs can be found in Figure 12.1. Because of these advantages, PMSMs have been widely used in electric vehicles, subway traction systems, wind power generation systems, etc. [78].

The study of how to improve the control performance of PMSMs is always a hot topic in the literature. Linear control schemes, e.g., proportional-integral (PI) control, is often used in PMSM control because of its simplicity. However, PI control is usually designed based on a linearized model, but PMSM is essentially a strongly coupled nonlinear system. Therefore, the performance of PI control cannot be guaranteed under some complex working conditions [23]. Besides, how to tune the PI parameters is also not a trivial task.

To solve the above problem, some researchers have studied the nonlinear control methods for PMSMs, such as sliding-mode control [104, 2, 85], passive control [108, 84], and backstepping control [120]. Among these methods, backstepping control divides the high-order PMSM system into several subsystems by introducing several virtual control laws, so that the actual controller can be designed in a recursive way [142]. However, the stability analysis of backstepping control is based on an accurate mathematical model, so it is invalid if system uncertainties exist. The sliding mode

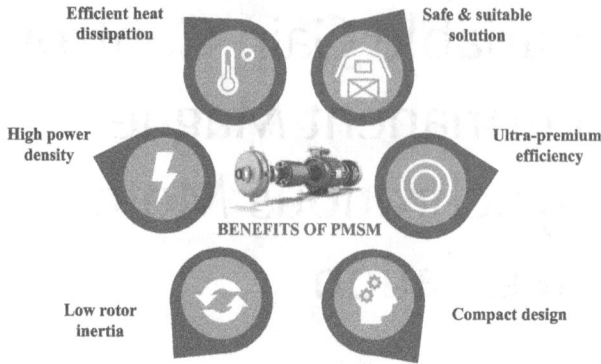

Figure 12.1 Benefits of PMSM.

control is able to deal with system uncertainties, but its inherent chattering problem is not acceptable for PMSM systems.

In practice, transient performance and steady-state performance are equally important for the control of PMSMs. As for transient performance, it is desirable that the designed controller ensures a fast convergence rate with a small overshoot [15]. From the perspective of convergence time, the above-mentioned nonlinear control methods can only guarantee asymptotic convergence, and are therefore not suitable for certain critical situations. The finite-time and fixed-time control methods can guarantee the desired control performance within a finite time, and are therefore able to meet the requirements for high-quality control on some special occasions [101]. However, in finite-time control, the convergence time depends on the system's initial states, resulting in a relatively long convergence time when the initial errors are large. In contrast, the convergence time of fixed-time control is independent of the initial states [143], so this chapter aims to achieve fixed-time control of PMSMs.

In particular, this chapter proposes a fixed-time control method for the speed regulation of PMSMs by using the dynamic gain control technique. The key feature of the proposed method is the design of two dynamic parameters. Specifically, a set of auxiliary variables is first introduced by a state transformation. These variables combine the original system states and the two introduced dynamic gains, thus facilitating the closed-loop system stability analysis. Then, these two dynamic gains are delicately designed by utilizing the Lyapunov method to ensure that all the speed tracking error tends to zero in a fixed time. Finally, several case studies are conducted under different working conditions to verify the advantages of the proposed fixed-time control strategy.

12.2 MATHEMATICAL MODEL AND PRELIMINARIES

For the convenience of controller design and stability analysis, some standard assumptions are given in the following, as those in [104, 67].

- The spatial magnetic field is sinusoidal distribution.
- The magnetic circuit is unsaturated.
- The effects of hysteresis and eddy current loss are neglected.

Under these assumptions, the mathematical model of the PMSM can be built as

$$
\begin{aligned}
\frac{d\omega}{dt} &= \frac{3n_p\Phi i_q}{2J} - \frac{B\omega}{J} - \frac{T_l}{J} \\
\frac{di_q}{dt} &= -\frac{R_s i_q}{L_q} - \frac{n_p\omega L_d i_d}{L_q} - \frac{n_p\omega\Phi}{L_q} + \frac{u_q}{L_q} \\
\frac{di_d}{dt} &= -\frac{R_s i_d}{L_d} + \frac{n_p\omega L_q i_q}{L_d} + \frac{u_d}{L_d}
\end{aligned}
\tag{12.1}
$$

in the synchronous rotating coordinate, i.e., the $d-q$ frame [78]. ω, i_q and i_d are the rotor angular speed, d-axes stator current, and q-axes stator current, respectively. The control input signals u_d and u_q are the stator winding voltages in the $d-q$ frame. n_p, Φ, B, T_l, J, R_s, L_q and L_d represent the pole-pair number, rotor flux, viscous friction coefficient, load torque, moment of inertia, stator resistance, d-axes stator inductance, and q-axes stator inductance, respectively.

For system (12.1), the control objectives are summarized as follows.

- *Objective I*: construct u_q such that ω tracks its reference ω^* in a fixed time;
- *Objective II*: construct u_d such that i_d tracks i_d^* in a fixed time.

12.3 FIXED-TIME CONTROL DESIGN

In this section, a fixed-time dynamic gain controller is constructed for fast and accurate speed control of PMSMs. First, the speed tracking errors are defined as:

$$
\begin{aligned}
x_1 &= \omega - \omega^* \\
x_2 &= i_q - \frac{2}{3n_p\Phi}B\omega^* - \frac{2}{3n_p\Phi}T_l
\end{aligned}
\tag{12.2}
$$

where T_l is usually a constant or slowly varying variable, implying that $\dot{T_l} = 0$.

Combining (12.1) and (12.2), one has

$$
\begin{aligned}
\frac{dx_1}{dt} &= \frac{3n_p\Phi}{2J}x_2 - \frac{B}{J}x_1 \\
\frac{dx_2}{dt} &= -\frac{R_s i_q}{L_q} - \frac{n_p\omega L_d i_d}{L_q} - \frac{n_p\omega\Phi}{L_q} + \frac{u_q}{L_q}.
\end{aligned}
\tag{12.3}
$$

Based on the variable gain control design method, we consider the following state transformation

$$
z_1 = \frac{x_1}{r_1^{2+\tau}r_2}, \quad z_2 = \frac{x_2}{r_1^{1+2\tau}r_2^2}
\tag{12.4}
$$

where $0 < \tau < 1$, and r_1 and r_2 are dynamic gains which will be designed later.

Based on (12.3), the derivative of (12.4) is computed as:

$$\frac{dz_1}{dt} = \frac{r_2}{r_1^{1-\tau}}\frac{3n_p\Phi}{2J}z_2 - \frac{B}{J}z_1 - \frac{\dot{r}_2}{r_2}z_1 - (2+\tau)\frac{\dot{r}_1}{r_1}z_1$$

$$\frac{dz_2}{dt} = \frac{1}{r_1^{1+2\tau}r_2^2}\left(-\frac{R_s i_q}{L_q} - \frac{n_p\omega L_d i_d}{L_q} - \frac{n_p\omega\Phi}{L_q} + \frac{u_q}{L_q}\right) - 2\frac{\dot{r}_2}{r_2}z_2 - (1+2\tau)\frac{\dot{r}_1}{r_1}z_2.$$

$$(12.5)$$

Then, for the speed control of PMSM, we construct the controller as:

$$u_q = L_q\left(-r_1^{3\tau}r_2^3(k_1z_1 + k_2z_2) + \frac{R_s i_q}{L_q} + \frac{n_p\omega L_d i_d}{L_q} + \frac{n_p\omega\Phi}{L_q}\right),\qquad(12.6)$$

where k_1 and k_2 are positive design constants.

Substituting (12.6) into (12.5), one has

$$\dot{z} = \frac{r_2}{r_1^{1-\tau}}Az - \frac{\dot{r}_2}{r_2}D_1z - \frac{\dot{r}_1}{r_1}D_2z + F\qquad(12.7)$$

where $z = [z_1, z_2]$ and

$$A = \begin{pmatrix} 0 & \frac{3n_p\Phi}{2J} \\ -k_1 & -k_2 \end{pmatrix}, D_1 = \begin{pmatrix} 1 & 0 \\ 0 & 2 \end{pmatrix}, D_2 = \begin{pmatrix} 2+\tau & 0 \\ 0 & 1+2\tau \end{pmatrix}, F = \begin{pmatrix} -\frac{B}{J}z_1 \\ 0 \end{pmatrix}.$$

If k_1 and k_2 are properly chosen, there must exist a positive definite matrix P such that

$$PA + A^T P \le -I,$$
$$\alpha_2 I \ge P \ge \alpha_1 I,$$
$$\beta_1 I \ge PD_1 + D_1 P \ge 0,$$
$$\beta_2 I \ge PD_2 + D_2 P \ge 0,$$

with α_1, α_2, β_1, β_2 being positive constants and I being an identity matrix.

Next, the dynamic parameters r_1, r_2 are designed as:

$$\frac{dr_1}{dt} = -\frac{1}{2\beta_1}r_1^\tau r_2 + \frac{r_2}{2\beta_1 r_1^{2-\tau}}\min\left\{\|z\|^2, 1\right\}, 0 < r_1(0) \le 1,$$

$$\frac{dr_2}{dt} = \frac{r_2^2}{4\alpha_2 r_1^{1-\tau}}\max\left\{\frac{\|z\|^{2\mu}}{r_2^{\mu+1}} - 1, 0\right\}, r_2(0) \ge \max\left\{8\|P\|\left|\frac{B}{J}\right|, 1\right\},$$

$$(12.8)$$

where $\mu > 0$ is a tuning constant.

Now we are ready to prove the stability of the closed-loop system. Choose the Lyapunov function as

$$V = z^T P z.$$

Then, the derivative of V is calculated as:

$$\dot{V}|_{(12.7)} = 2z^T P\dot{z}$$

$$\le -\frac{r_2}{r_1^{1-\tau}}\|z\|^2 - \frac{\dot{r}_2}{r_2}z^T(PD_1 + D_1P)z$$

$$- \frac{\dot{r}_1}{r_1}z^T(PD_2 + D_2P)z + 2z^T PF.$$

Since $\dot{r}_2 \geq 0$, $r_2 \geq r_2(0)$, and $\dot{r}_1 \geq -\frac{1}{2\beta_1}r_1^\tau r_2$, one has

$$\dot{V}|_{(12.7)} \leq -\frac{r_2}{r_1^{1-\tau}}\|z\|^2 + \frac{r_2}{2\beta_1 r_1^{1-\tau}}z^T(PD_2 + D_2 P)z + 2z^T PF$$

$$\leq -\frac{r_2}{r_1^{1-\tau}}\|z\|^2 + \frac{r_2}{2r_1^{1-\tau}}\|z\|^2 + 2z^T PF$$

$$\leq -\frac{r_2}{2r_1^{1-\tau}}\|z\|^2 + 2z^T PF.$$

Using the dynamic gain laws (12.8) and Young's inequality, one has

$$2z^T PF \leq 2\left|\frac{B}{J}\right|\,\|P\|\,\|z\|^2 \leq \frac{r_2}{4r_1^{1-\tau}}\|z\|^2.$$

Therefore, $\dot{V}$ is rewritten as:

$$\dot{V}|_{(12.7)} \leq -\frac{r_2}{4r_1^{1-\tau}}\|z\|^2.$$

Then, the fixed-time stability of the closed-loop system can be obtained by following the following three steps.

Step 1: Prove $\|z\|^2 \leq r_2$ within a fixed time.

Defining a positive function $V_1 = \frac{1}{r_2}V$, then its derivative is given as:

$$\dot{V}_1 \leq -\frac{1}{4r_1^{1-\tau}}\|z\|^2 - \frac{\dot{r}_2}{r_2^2}V \leq -\frac{1}{4\alpha_2 r_1^{1-\tau}}V_1\frac{\|z\|^{2\mu}}{r_2^\mu} \leq -\frac{1}{4\alpha_2^{1+\mu}}V_1^{1+\mu}.$$

Thus, one has

$$V_1(t) \leq \left(\frac{1}{V_1^{-\mu}(0) + \frac{\mu}{4\alpha_2^{1+\mu}}t}\right)^{\frac{1}{\mu}}, \quad V_1(0) \neq 0.$$

Therefore, there exists a time instant t_1 such that $V_1(t) \leq \alpha_1$ when $t > t_1$. Since $V_1(t) \geq \alpha_1 \frac{1}{r_2}\|z\|^2$, we have $\|z\|^2 \leq r_2$.

Step 2: Prove $\|z\|^2 \leq 1$ within a fixed time.

When $\|z\|^2 \leq r_2$, one has

$$\dot{V} \leq -\frac{1}{4r_1^{1-\tau}}\|z\|^4 \leq -\frac{1}{4}\frac{1}{\alpha_2^2}V^2.$$

Then, it holds that

$$V(t) \leq \frac{1}{\frac{1}{V(t_0)} + \frac{1}{4\alpha_2^2}(t - t_0)}, \quad t \geq t_0.$$

Therefore, there exists a time instant t_2 such that $V \leq \alpha_1$ when $t > t_2$. Due to $V(t) \geq \alpha_1 \|z(t)\|^2$, we obtain that $\|z\|^2 \leq 1$.

Step 3: Prove the fixed-time stability of the closed-loop system.

Let

$$V_2 = V + \frac{1}{8}\beta_1 r_1^2.$$

The derivative of V_2 is given as:

$$\dot{V}_2 \leq -\frac{r_2}{4r_1^{1-\tau}}\|z\|^2 - \frac{1}{8}r_1^{1+\tau}r_2 + \frac{r_2}{8r_1^{1-\tau}}\|z\|^2$$

$$\leq -\frac{r_2}{8r_1^{1-\tau}}\|z\|^2 - \frac{1}{8}r_1^{1+\tau}r_2$$

$$\leq -\frac{1}{8r_1^{1-\tau}}\|z\|^2 - \frac{1}{8}r_1^{1+\tau}$$

$$\leq -\frac{2}{8(1+\tau)}\|z\|^{1+\tau} + \frac{1-\tau}{8(1+\tau)}r_1^{1+\tau} - \frac{1}{8}r_1^{1+\tau}$$

$$\leq -\frac{2}{8(1+\tau)}\|z\|^{1+\tau} - \frac{2\tau}{8(1+\tau)}r_1^{1+\tau}$$

$$\leq -\bar{\omega}V_2^{\frac{1+\tau}{2}}.$$

where $\bar{\omega} = \min\{\frac{2}{8(1+\tau)}, \frac{2\tau}{8(1+\tau)}\}$. Using the finite-time Lyapunov stability criterion [3], V_2 converges to zero within a finite time and $V_2(t_2) \leq \alpha_2 + \frac{1}{8}\beta_1$. Therefore, V_2 tends to zero in a fixed time. The calculation of the setting time has been discussed in Chapter 4.

Up to now, it has been proven that fixed-time speed tracking control is achieved. In order to avoid the singularity problem caused by r_1, the user can set a lower limit $\bar{r}_1 > 0$ for r_1 when implementing the proposed controller. This approach somewhat sacrifices the control performance, i.e., the speed regulation error converges not to zero but to a small neighborhood around zero.

The following presents the design procedure of the d-axes current controller. Define the d-axes current tracking error as

$$z_3 = i_d - i_d^*$$

where i_d^* is set to be zero to achieve the decoupling between the speed and the d-axes current.

Choose the Lyapunov function as

$$V_3 = \frac{1}{2}z_3^2.$$

The derivative of V_3 is given by

$$\dot{V}_3 = z_3\dot{z}_3 = z_3\left(-\frac{R_s i_d}{L_d} + \frac{n_p\omega L_q i_q}{L_d} + \frac{u_d}{L_d}\right). \tag{12.9}$$

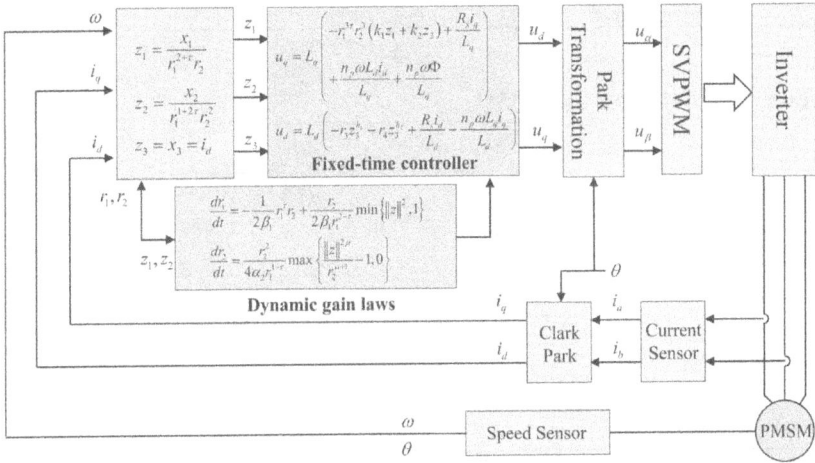

Figure 12.2 Diagram of the proposed fixed-time dynamic gain control for speed regulation of the PMSM.

Construct the d-axes current controller as

$$u_d = L_d \left(-r_3 z_3^{h_1} - r_4 z_3^{h_2} + \frac{R_s i_d}{L_d} - \frac{n_p \omega L_q i_q}{L_d} \right) \tag{12.10}$$

where $h_1 = 5/7$ and $h_2 = 5/3$.

Substituting (12.10) into (12.9) yields

$$\dot{V}_3 = -r_3 z_3^{h_1+1} - r_4 z_3^{h_2+1}$$
$$= -r_3 \left(z_3^2 \right)^{\frac{h_1+1}{2}} - r_4 \left(z_3^2 \right)^{\frac{h_2+1}{2}}$$
$$= -2^{\frac{h_1+1}{2}} r_3 V_3^{\frac{h_1+1}{2}} - 2^{\frac{h_2+1}{2}} r_4 V_3^{\frac{h_2+1}{2}}.$$

Since (12.3) meets Lemma 1 in [3], the closed-loop current control system is fixed-time stable. Therefore, z_3 tends to zero in a fixed time.

Figure 12.2 shows the block diagram of the proposed control method. The overall system consists of a PMSM, a space vector pulse-width-modulation (SVPWM), a voltage-source inverter, measurement sensors, and the proposed controller (including the state transformation, fixed-time controllers, and dynamic gain laws). As can be seen, the proposed control method has only a single-loop control structure, and is therefore easy to implement. The main results of this chapter are summarized below.

Table 12.1

System parameters of the PMSM.

Description	Symbol	Value	Unit
Rated speed	*	1000	r/min
Rated load torque	T_l	3	N·m
Pole-pair number	n_p	4	*
Stator resistance	R_s	0.93	Ω
d-axes inductance	L_q	3	mH
q-axes inductance	L_d	3	mH
Rotor flux	Φ	0.29	Wb
Moment of inertia	J	0.003	kg·m^2

Theorem 12.1

For the considered PMSM system (12.1), by using the dynamic state transformation (12.4), the controllers (12.6) and (12.10), and the dynamic gain law (12.8), the control objectives I-II are both achieved. ∎

12.4 CASE STUDIES

To verify the effectiveness of the proposed fixed-time control method for PMSMs, several case studies are conducted in MATLAB/Simulink. The PMSM parameters are shown in Table 12.1. Meanwhile, the conventional PI control is also carried out and compared with the proposed method. In order to make a fair comparison, the selection of control parameters follows two principles: 1) the parameters are able to achieve satisfactory control performance, and 2) the control efforts are at the same level.

12.4.1 CASE 1: START-UP PERFORMANCE

The speed reference ω^* is 1000 r/min and the load torque T_l is 3 N·m. Figures 12.3–12.6 show the results when the PMSM starts up. The speed response curves and tracking error are given in Figure 12.3 and Figure 12.4, respectively. Although the speed is regulated to 1000 r/min for both methods, the transient response is quite different. The settling time of the rotor speed is 42 ms for the PI control and 13 ms for the proposed fixed-time control method. Figure 12.6 shows the response curves of d-axes current, which shows that these two methods both achieve satisfactory current control performances.

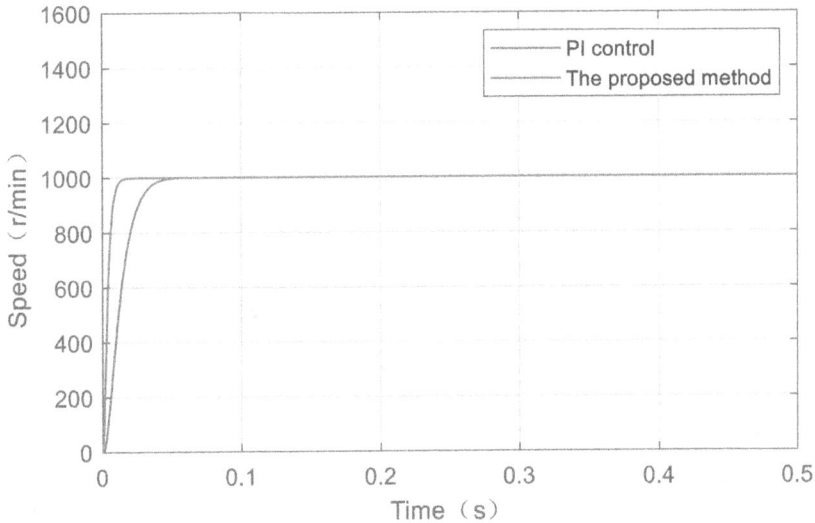

Figure 12.3 Curves of the speed when the PMSM starts up.

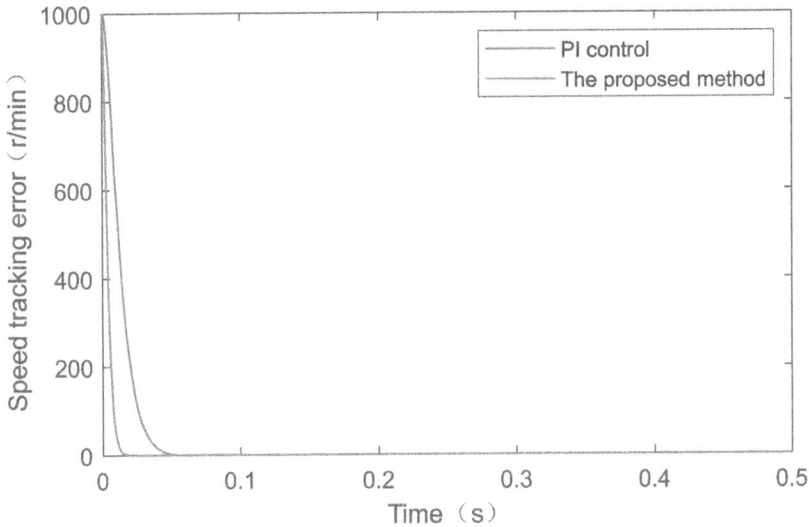

Figure 12.4 Curves of the speed tracking error when the PMSM starts up.

12.4.2 CASE 2: DIFFERENT WORKING CONDITIONS

In this test, the load torque changes from 3 N·m to 5 N·m at $t = 0.5$ s. Figures 12.7–12.8 give the response curves of the speed. When the load changes, the rotor speed drops by about 8 r/min for the PI control method, and the recovering time is

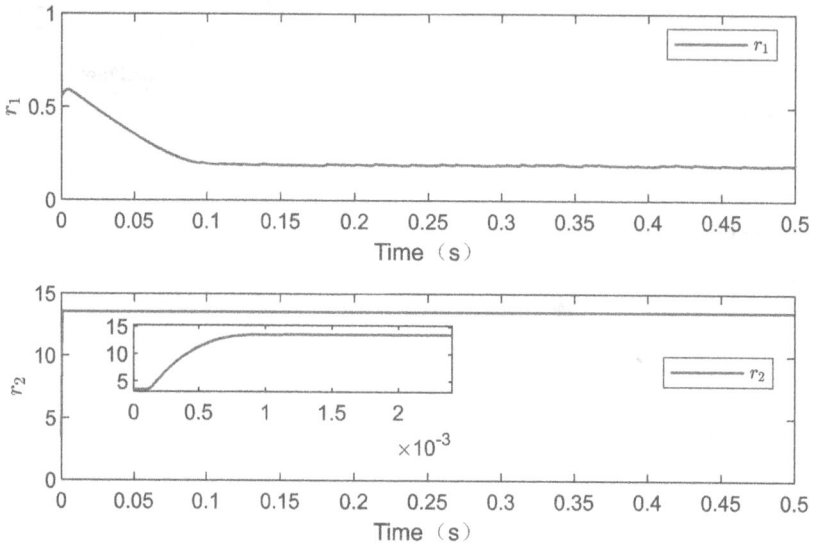

Figure 12.5   Curves of dynamic gains r_1 and r_2 when the PMSM starts up.

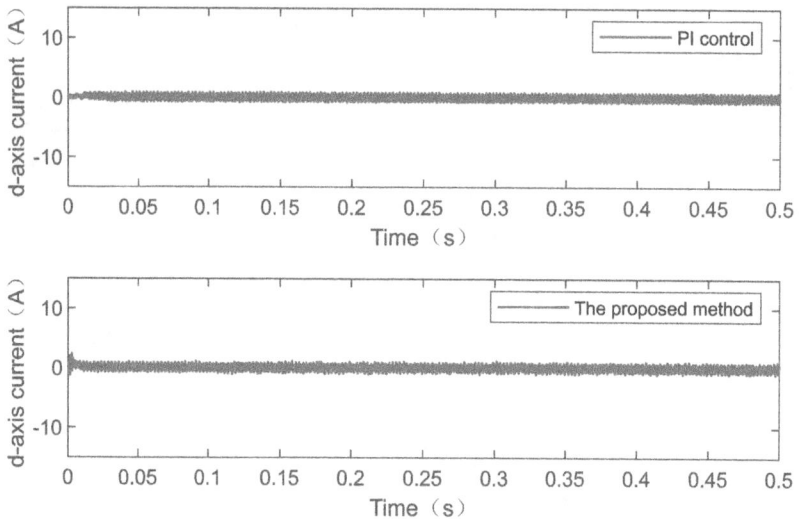

Figure 12.6   Curves of the d-axes current when the PMSM starts up.

52 ms. For the proposed fixed-time control method, the load variation only has a minor effect on the rotor speed. This indicates that the proposed method is robust against load changes. Moreover, when the speed reference changes from 1000 r/min

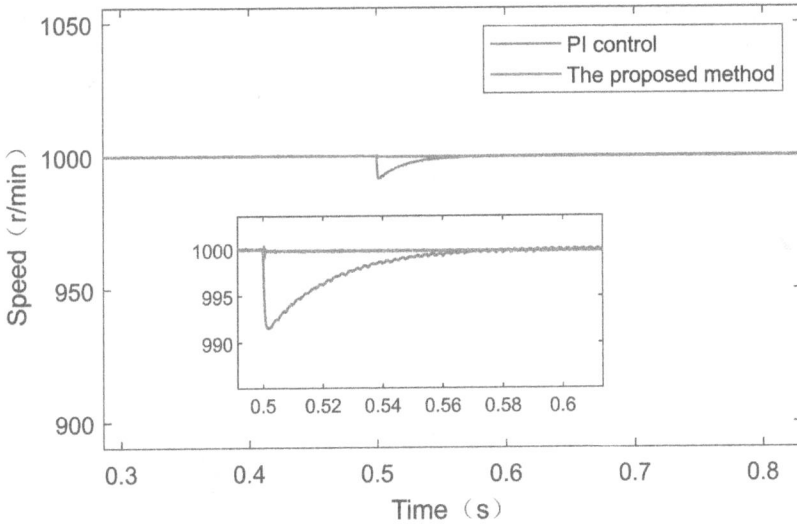

Figure 12.7   Curves of the speed when the load changes.

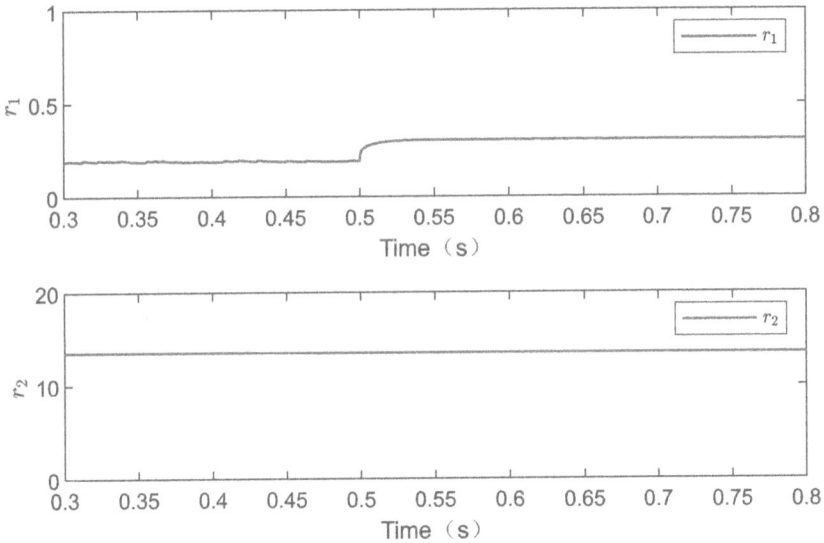

Figure 12.8   Curves of dynamic gains r_1 and r_2 when the load changes.

to 1300 r/min at $t = 1$ s, as shown in Figures 12.9–12.10, the proposed method performs faster speed tracking than the PI method. In summary, the proposed method is able to adjust the control gains according to different working conditions, ensuring faster and more accurate speed control than the conventional PI method.

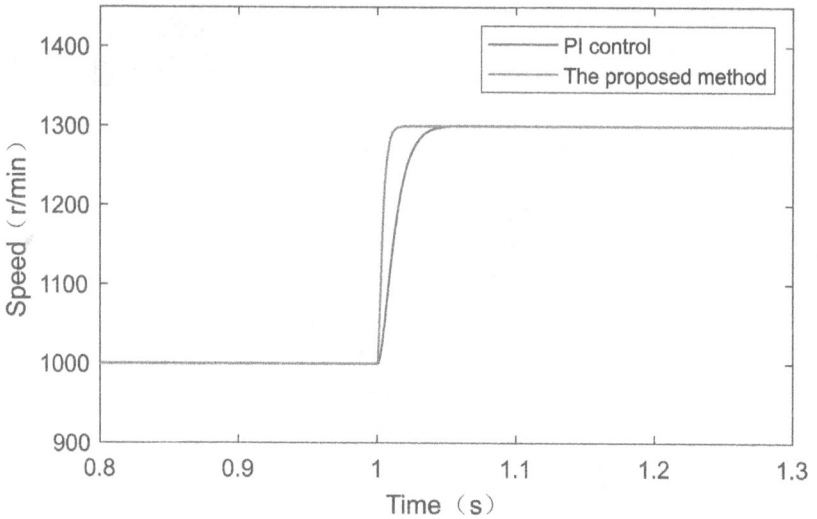

Figure 12.9   Curves of the speed when the reference changes.

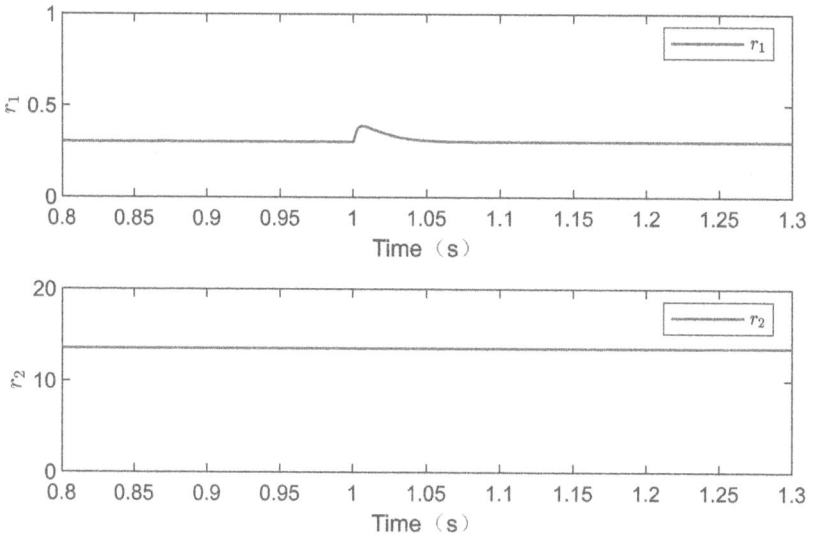

Figure 12.10   Curves of the dynamic gains r_1 and r_2 when the reference changes.

12.5   NOTES

In this chapter, a fixed-time dynamic gain control method for PMSM speed regulation is proposed. The controller design is transformed into the control gain design by introducing the state transformation with two dynamic gains. Then, these two dynamic gains are sensitively designed by utilizing the Lyapunov method to ensure that the speed tracking error tends to zero in a fixed time. Case studies with different working conditions are conducted to verify the effectiveness of the proposed method. Compared with the traditional PI method, the proposed fixed-time method presents faster and more accurate speed tracking performance.

13 Distributed Robust Secondary Control of Islanded Microgrids

In order to solve the voltage and frequency deviation problem caused by primary control, this chapter proposes a distributed robust secondary control method with a static gain for an inverter-based islanded microgrid (MG). To facilitate improved controller design, the large-signal mathematical model of the MG is first transformed into two simple subsystems with a strict-feedback structure. Then, the system uncertainties are grouped together as a lumped disturbance, and a super-twisting disturbance observer is proposed to estimate this lumped disturbance. Finally, the distributed secondary controller is constructed by using the static gain control approach. Case studies verify the effectiveness of the proposed distributed secondary control method.

13.1 BACKGROUND

Driven by environmental concerns, renewable energy has been increasingly integrated into modern power systems to gradually replace traditional fossil energy [165]. Microgrids (MGs) provide an effective way to integrate distributed renewable energy, and how to manage the distributed generators (DGs) in an MG has attracted significant attention. An MG can operate in either the grid-connected mode or islanded mode, as shown in Figure 13.1. In the islanded mode, the control strategy needs to be carefully designed as it directly affects the operation performance of the MG system. A series of primary control strategies, such as droop control, has been widely used to regulate the output voltage and frequency. However, these primary control methods inevitably deviate the voltage and frequency from their nominal values. Therefore, another control layer, i.e., the secondary control, is needed to compensate for these deviations [52].

The conventional centralized secondary control strategy requires a complex communication network, which may adversely affect the system's reliability and decrease the plug-and-play capability. To tackle this problem, distributed secondary control with a sparse communication network is proposed. In [4, 5], a distributed cooperative secondary control strategy is proposed. Even though only a few DGs in the MG have access to the reference frequency and voltage values, the frequency and voltage can be well regulated by allowing each DG to communicate with their neighbors. In [64], the authors applied sliding-mode control and backstepping control methods to improve secondary voltage control performance. In [73], the distributed secondary control in discrete time is discussed. In [36], finite-time voltage control is proposed to ensure voltage restoration within a finite time. Although many in-depth studies

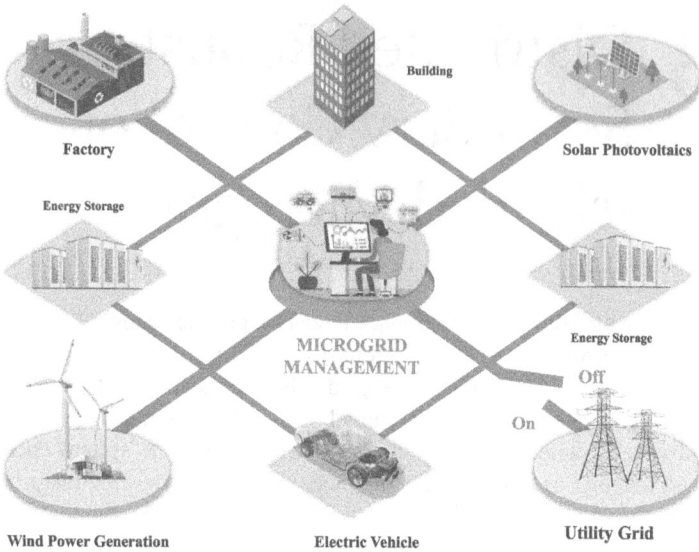

Figure 13.1   Intelligent microgrids.

on distributed secondary control have been conducted, most of them are developed based on accurate DG models without considering uncertainties and require a large number of system states to be available for controller design. This requirement is infeasible in practice because of the limited deployment of measurement devices [29]. In [19] and [98], the super-twisting and full-order sliding-mode control strategies are proposed for distributed secondary control to enhance system robustness. However, they both require prior knowledge of the upper bounds of uncertainties' derivatives, which is normally unavailable for a practical microgrid system.

This chapter introduces a distributed robust secondary control scheme via the static gain control approach. Firstly, the MG's complex large-signal model is transformed into two single-input single-output models. Then, a super-twisting disturbance observer is designed to estimate the lumped disturbance, thereby improving the system robustness and saving the cost of sensor configuration. Theoretical analysis and simulation studies verify the effectiveness of the proposed distributed secondary control method.

13.2   MATHEMATICAL MODEL AND PRELIMINARIES

COMMUNICATION NETWORK

Consider N DGs sharing an undirected communication graph $\mathscr{G} = (\mathscr{V}, \mathscr{E})$. The set $\mathscr{V} = \{1, 2, \ldots, N\}$ and $\mathscr{E} \subseteq \mathscr{V} \times \mathscr{V}$ represent the node set and edge set, respectively. $(i, j) \in \mathscr{E}$ denotes the edge between node i and j. The weighted adjacent matrix $\mathscr{A} = (a_{ij}) \in \mathbb{R}^{N \times N}$ with $a_{ij} = a_{ji} = 1$ for $(i, j) \in \mathscr{E}$, $a_{ij} = a_{ji} = 0$ for $(i, j) \notin \mathscr{E}$.

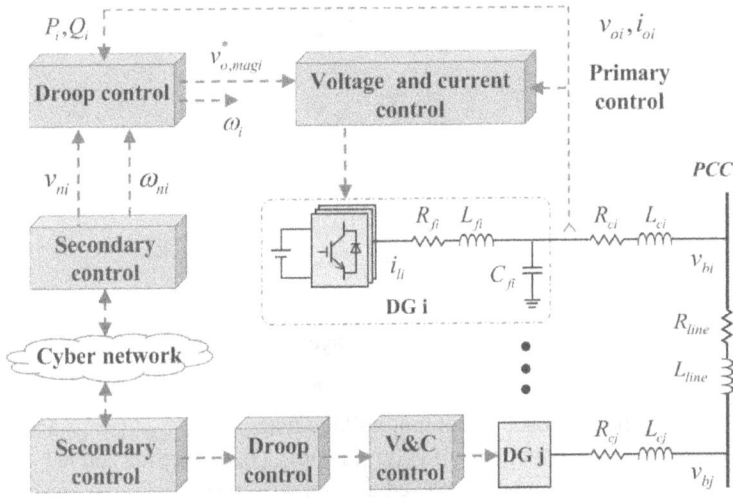

Figure 13.2 The physical structure of an islanded microgrid with inverter-based DGs.

It is assumed $a_{ii} = 0$ for any $i \in \mathcal{V}$. The Laplacian matrix is $\mathscr{L} = \mathscr{D} - \mathscr{A}$, where $\mathscr{D} = \text{diag}\{d_1, d_2, \ldots, d_N\}$ with $d_i = \sum_{j=1}^{N} a_{ij}$. In this chapter, the communication network among DGs is assumed to be undirected and connected, i.e., there exists a path between any two nodes. Assume there exists a virtual leader (agent 0), and define the leader adjacent matrix as $\mathscr{B} = \text{diag}\{b_1, b_2, \ldots, b_N\}$, in which $b_i = 1$ if there is an edge from the node 0 to node i, otherwise, $b_i = 0$.

MATHEMATICAL MODEL

Figure 13.2 shows the physical structure of an islanded microgrid with inverter-based DGs. Each DG consists of the prime dc source (e.g., photovoltaic panels or fuel cells), a dc/ac inverter, a LC filter, and an RL output connector. These inverters operate in the voltage control mode when the MG is in the islanded mode. The dynamics of the ith DG are formulated on the (d-q) frame at a rotating frequency ω_i. The primary control layer includes the outer droop control, and the inner voltage and current control. δ_i denotes the rotating angle difference of the ith DG and satisfies

$$\dot{\delta}_i = \omega_i - \omega_{com}.$$

The droop control balances the power supply and demand by adjusting the frequency and the output voltage amplitude of the ith DG. The frequency and voltage droop control is described as

$$\omega_i = \omega_{ni} - m_{P_i} P_i,$$
$$v_{odi}^* = v_{o,magi}^* = v_{ni} - n_{Q_i} Q_i,$$
$$v_{oqi}^* = 0,$$

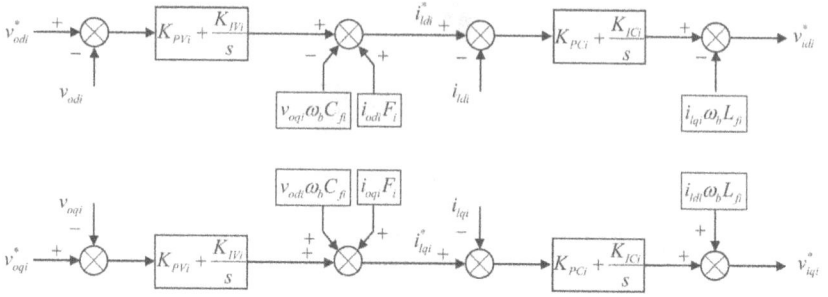

Figure 13.3　The structure of the inner voltage and current control.

where $v^*_{o,magi}$ is the reference for output voltage magnitude $v_{o,magi}$; v^*_{odi} and v^*_{oqi} are the d-axis and q-axis components of $v^*_{o,magi}$; P_i, m_{P_i}, Q_i, n_{Q_i} are the active power, frequency droop coefficient, reactive power, and voltage droop coefficient, respectively; and ω_{ni} and v_{ni} are primary control references generated by the secondary control.

The diagram block of voltage and current control is given in Figure 13.3, where the dynamic model is given by

$$\dot{\phi}_{di} = v^*_{odi} - v_{odi}$$

$$\dot{\phi}_{qi} = v^*_{oqi} - v_{oqi}$$

$$i^*_{ldi} = F_i i_{odi} - \omega_b C_{fi} v_{odi} + K_{PVi}\left(v^*_{odi} - v_{odi}\right) + K_{IVi}\phi_{di}$$

$$i^*_{lqi} = F_i i_{oqi} + \omega_b C_{fi} v_{odi} + K_{PVi}\left(v^*_{oqi} - v_{oqi}\right) + K_{IVi}\phi_{qi}$$

$$\dot{\gamma}_{di} = i^*_{ldi} - i_{ldi}$$

$$\dot{\gamma}_{qi} = i^*_{lqi} - i_{lqi}$$

$$v^*_{idi} = -\omega_b L_{fi} i_{ldi} + K_{PCi}\left(i^*_{ldi} - i_{ldi}\right) + K_{ICi}\gamma_{di}$$

$$v^*_{iqi} = \omega_b L_{fi} i_{ldi} + K_{PCi}\left(i^*_{lqi} - i_{lqi}\right) + K_{ICi}\gamma_{qi}$$

where ϕ_{di}, ϕ_{qi}, γ_{di}, γ_{di} are auxiliary states of the PI controllers; ω_b is the rated angular frequency; K_{PVi}, K_{IVi}, K_{PCi}, K_{ICi} are PI control parameters; i_{odi} and i_{oqi} are the direct and quadrature components of output current; and i_{ldi} and i_{lqi} are the direct and quadrature components of i_{li}.

Based on [4, 166], the nonlinear large-signal nonlinear model of the ith DG is

$$\dot{\chi}_i = F_i(\chi_i) + K_i(\chi_i) D_i + G_i(\chi_i) u_i$$

$$y_i = (v_{odi},\ \omega_i)^T \tag{13.1}$$

where $\chi_i = \left(\delta_i\ P_i\ Q_i\ \phi_{di}\ \phi_{qi}\ \gamma_{di}\ \gamma_{qi}\ i_{ldi}\ i_{lqi}\ v_{odi}\ v_{oqi}\ i_{odi}\ i_{oqi}\right)^T$. $u_i = (v_{ni},\ \omega_{ni})^T$ is the system input and y_i is the system output. Detailed definitions and expressions of F_i, K_i, D_i, G_i have been given in [4, 166] and are omitted here for brevity.

The control objectives of the secondary control are to design v_{ni} and ω_{ni} which can ensure

- *Objective I*: v_{odi} tracks its rated value v_{ref};
- *Objective II*: ω_i tracks its rated value ω_{ref} while achieving desired active power sharing, i.e., $m_{P_i} P_i = m_{P_j} P_j$ for $i, j \in \mathcal{V}$.

The rated values v_{ref} and ω_{ref} are provided by the leader node.

13.3   DISTRIBUTED SECONDARY VOLTAGE CONTROL

By using the input-output partial feedback linearization technique, the dynamic model (13.1) can be transformed into the following second-order system:

$$\begin{aligned} \dot{x}_{i,1} &= x_{i,2} \\ \dot{x}_{i,2} &= f_i(\chi_i) + g_i v_{ni} \end{aligned} \tag{13.2}$$

where

$$
\begin{aligned}
f_i &= \left(-\omega_i^2 - \frac{K_{Pvi}K_{Pci}+1}{C_{fi}L_{fi}} - \frac{1}{C_{fi}L_{ci}} \right) v_{odi} - \frac{\omega_b K_{Pci}}{L_{fi}} v_{oqi} \\
&\quad + \frac{R_{ci}}{C_{fi}L_{ci}} i_{odi} - \frac{2\omega_i}{C_{fi}} i_{oqi} - \frac{R_{fi}+K_{Pci}}{C_{fi}L_{fi}} i_{ldi} \\
&\quad + \frac{2\omega_i - \omega_b}{C_{fi}} i_{lqi} - \frac{K_{Pvi}K_{Pci}}{C_{fi}L_{fi}} n_{Qi} Q_i + \frac{K_{Ivi}K_{Pci}}{C_{fi}L_{fi}} \phi_{di} \\
&\quad + \frac{K_{Ici}}{C_{fi}L_{fi}} \gamma_{di} + \frac{1}{C_{fi}L_{ci}} v_{bdi} + \zeta_i \\
g_i &= \frac{K_{Pvi}K_{Pci}}{C_{fi}L_{fi}}, \quad x_{i,2} = \omega_i v_{oqi} + \frac{i_{ldi} - i_{odi}}{C_{fi}}
\end{aligned}
$$

where $x_{i,1} = v_{odi}$ and v_{bdi} is the q-axis voltage at the connection bus between the ith DG and microgrid [30]. ζ_i represents model uncertainties, including external disturbances, the coupling between voltage and frequency control, etc.

For a practical microgrid, the values of C_{fi} and L_{fi} may vary with the environment. These factors make f_i uncertain, bringing difficulties to the controller design. To solve this problem and enhance robustness, this article considers f_i as a lumped disturbance and assumes $|\dot{f}_i| \leq \bar{\omega}$, where $\bar{\omega}$ is an unknown constant.

Then, a super-twisting disturbance observer (STDO) is designed as

$$
\begin{cases}
\dot{\hat{x}}_{i,2} = \hat{f}_i + g_i v_{ni} \\
\hat{f}_i = -k_{d1} |\tilde{x}_{i,2}| \text{sign}(\tilde{x}_{i,2}) - k_{d2} \int_{t_0}^{t} \text{sign}(\tilde{x}_{i,2}) ds
\end{cases}
\tag{13.3}
$$

to estimate the lumped disturbances, where $\hat{f}_i$ is the estimation of f_i, $\tilde{x}_{i,2} = \hat{x}_{i,2} - x_{i,2}$, and k_{d1}, k_{d2} are positive constants.

Using (13.2) and (13.3), it is obtained that

$$\dot{\tilde{x}}_{i,2} = \hat{f}_i - f_i.$$

It can be observed if $x_{i,2} = \dot{x}_{i,2} = 0$, then $\hat{f}_i = f_i$. According to (16)-(18) in [140], one can select proper $k_{d1} > 0$ and $k_{d2} > 0$ to guarantee $x_{i,2} = \dot{x}_{i,2} = 0$ after a finite-time transient.

The following gives the design of the distributed voltage controller using the static gain control approach. The distributed control protocol is constructed as

$$v_{ni} = -g_i^{-1} \left(L^3 k_1 \mathscr{L} \left(\sum_{j=1}^{N} a_{ij}(x_i - x_j) + b_i(x_i - x_0) \right) + \hat{f}_i \right), \tag{13.4}$$

where $x_0 \triangleq (x_{0,1}, x_{0,2})^T = (v_{ref}, 0)^T$, and $\mathscr{L} = \mathrm{diag}\{\frac{1}{L}, \frac{1}{L^2}\}$ with $L \geq 1$ being a constant parameter.

Firstly, we consider the state transformation

$$z_{i,1} = \frac{1}{L}(x_{i,1} - x_{0,1}), \quad z_{i,2} = \frac{1}{L^2}(x_{i,2} - x_{0,2}).$$

From the dynamic system (13.2), we obtain

$$\begin{cases} \dot{z}_{i,1} = L z_{i,2} \\ \dot{z}_{i,2} = \dfrac{1}{L^2}(f_i(\chi_i) + g_i v_{ni}) \end{cases} \tag{13.5}$$

and the distributed control protocol (13.4) is converted into

$$v_{ni} = -g_i^{-1} \left(L^3 k_1 \left(\sum_{j=1}^{N} a_{ij}(z_i - z_j) + b_i z_i \right) + \hat{f}_i \right) \tag{13.6}$$

where $z_i = (z_{i,1}, z_{i,2})^T$, and $k_1 = (k_{11}, k_{12})$ with k_{11} and k_{12} being constant to be determined.

Substituting (13.6) into (13.5), it can be obtained that

$$\begin{cases} \dot{z}_{i,1} = L z_{i,2} \\ \dot{z}_{i,2} = -L k_1 \left(\displaystyle\sum_{j=1}^{N} a_{ij}(z_i - z_j) + b_i z_i \right) + \dfrac{1}{L^2}(f_i(\chi_i) - \hat{f}_i). \end{cases} \tag{13.7}$$

It follows from the analysis of the STDO that $\hat{f}_i = f_i(\chi_i)$ after a finite-time transient. Then, (13.7) is rewritten as

$$\begin{cases} \dot{z}_{i,1} = L z_{i,2} \\ \dot{z}_{i,2} = -L k_1 \left(\displaystyle\sum_{j=1}^{N} a_{ij}(z_i - z_j) + b_i z_i \right) \end{cases}$$

when $t \geq T$ with T being a positive constant.

Denoting $z = (z_1^T, \ldots, z_N^T)^T$, we get the compact form as

$$\dot{z} = L(I_N \otimes A + \mathscr{L} \otimes B k_1)z \tag{13.8}$$

where $A = (a_{ij})_{2\times 2}$ with $a_{12} = 1$ and $a_{11} = a_{21} = a_{22} = 0$, $B = (0\ 1)^T$, $\mathcal{L}$ is the Laplacian matrix of the communication topology.

Following the result in Chapter 9, we can find the constants k_{11}, k_{12} such that $(I_N \otimes A + \mathcal{L} \otimes Bk_1)$ is the Hurwitz matrix. Then, a positive definite matrix P can be found to meet $P(I_N \otimes A + \mathcal{L} \otimes Bk_1) + (I_N \otimes A + \mathcal{L} \otimes Bk_1)^T P \leq -I$. Let $V = z^T P z$, and its derivative is computed as

$$\dot{V}_{(13.8)} \leq -L\|z\|^2.$$

Therefore, z exponentially converges to the origin, i.e., the voltage tracking error $z_i - z_0$ converges to the origin exponentially. The above analysis can be summarized in the following theorem.

Theorem 13.1

For the system (13.2), by using the state transformation (13.5), the distributed voltage controller (13.6) and the disturbance observer (13.3), the control objective I is achieved. ∎

13.4 DISTRIBUTED FREQUENCY CONTROL AND ACTIVE POWER SHARING

Similar to the above section, the nonlinear model (13.1) can be simplified as

$$\dot{\omega}_i = \dot{\omega}_{ni} - m_{P_i}\dot{P}_i = u_{i,\omega}$$

for the frequency control of the ith DG, where $u_{i,\omega}$ is the auxiliary frequency control input. Meanwhile, the active power sharing control system is

$$m_{P_i}\dot{P} = u_{i,P},$$

where $u_{i,P}$ is the auxiliary active power control input [5].

To achieve frequency regulation and active power sharing, the local errors are defined as

$$\begin{cases} e_{i,\omega} = \sum_{j=1}^{N} a_{ij}(\omega_i - \omega_j) + b_i(\omega_i - \omega_{ref}), \\ e_{i,P} = \sum_{j=1}^{N} a_{ij}(m_{P_i}P_i - m_{P_j}P_j), \qquad i, j \in \mathcal{V} \end{cases} \tag{13.9}$$

for the ith DG. Then, the auxiliary controllers are constructed as

$$\begin{cases} u_{i,\omega} = -(d_i + b_i)^{-1} k_2 e_{i,\omega}, \\ u_{i,P} = -d_i^{-1} k_2 e_{i,P}, \end{cases} \tag{13.10}$$

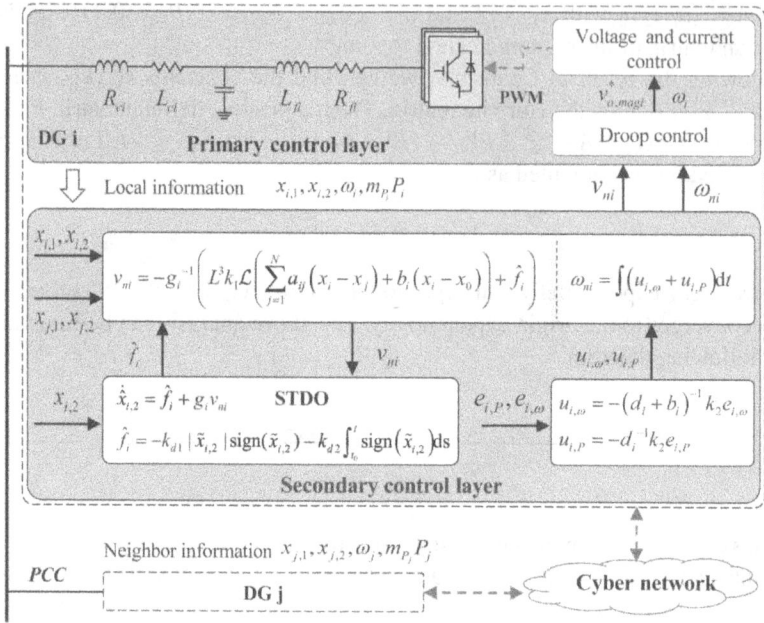

Figure 13.4 Control diagram of the proposed distributed secondary control method.

where k_2 is a positive constant. Then the primary frequency reference ω_{ni} is calculated by

$$\omega_{ni} = \int (u_{i,\omega} + u_{i,P}) dt.$$

Using the constructed controllers (13.10), the frequency tracking error $\omega_i - \omega_{ref}$ converges to the origin with a fast speed while achieving $m_{P_i} P_i = m_{P_j} P_j$, i.e., the control objective II is achieved, $i, j \in \mathcal{V}$. Since the stability analysis is easily obtained, the proof is omitted here. Figure 13.4 shows the control diagram of the proposed method.

13.5　CASE STUDIES

In this section, an islanded microgrid is built in MATLAB/SimPowerSystems to verify the effectiveness of the proposed method. The nominal output voltage and frequency are set as $v_{ref} = 311$ V and $\omega_{ref} = 50$ Hz. As shown in Figure 13.5, the microgrid consists of four DGs, four local loads, and three transmission lines. Figure 13.5 also gives the communication graph for the four DGs. For the weighted adjacent matrix, set $a_{ij} = 1$ for $(i, j) \in \mathcal{E}$. The leader adjacent matrix $B = \text{diag}\{1, 0, 0, 0\}$, i.e., only DG #1 accesses the reference values. The other system and control parameters are given in Table 13.1.

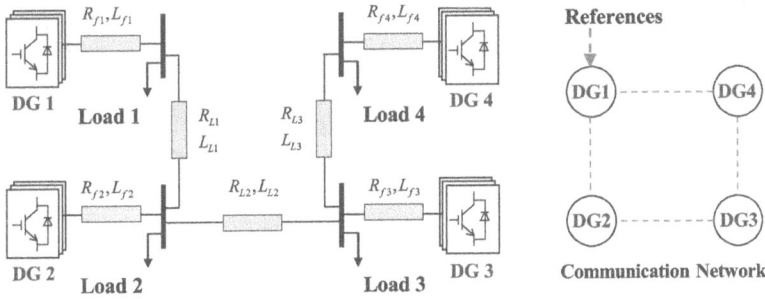

Figure 13.5 Structure of the tested islanded microgrid.

Table 13.1
System parameters configuration for the microgrid.

Parameters	DG 1	DG 2	DG3 & DG 4
$R_{fi}(\Omega)$	0.1	0.1	0.1
L_{fi}(mH)	1.35	1.35	1.35
$C_{fi}(\mu F)$	47	47	47
$R_{ci}(\Omega)$	0.02	0.02	0.04
L_{ci}(mH)	2	2	2
$m_{P_i}(\times 10^{-5})$	6.28	9.42	12.56
$n_{Q_i}(\times 10^{-4})$	5	7.5	10
K_{Pvi}	0.3	0.3	0.3
K_{Ivi}	390	390	420
K_{Pci}	3	3	3
K_{Ici}	16000	16000	20000
$R_{Li}(\Omega)$	0.23	0.35	0.23
L_{Li}(mH)	0.318	1.847	0.318
Load 1	$R_{Load1}=2\ \Omega,\ L_{Load1}=6.4$ mH		
Load 2	$R_{Load2}=4\ \Omega,\ L_{Load2}=9.6$ mH		
Load 3	$R_{Load3}=6\ \Omega,\ L_{Load3}=12.8$ mH		
Load 4	$R_{Load4}=6\ \Omega,\ L_{Load4}=12.8$ mH		

13.5.1   CASE 1: LOAD DISTURBANCES

In this test, an additional load is added to the microgrid to test the robustness of the proposed method against load disturbances. The test process is defined as follows.

1. $t = 0 \sim 1$ s, only the primary control works;
2. $t = 1$ s, the proposed secondary control is inserted;
3. $t = 3$ s, add a load $S_{Load5} = 12$ kW$+j15$ kVar to Load 3;
4. $t = 6$ s, remove the load S_{Load5} from Load 3.

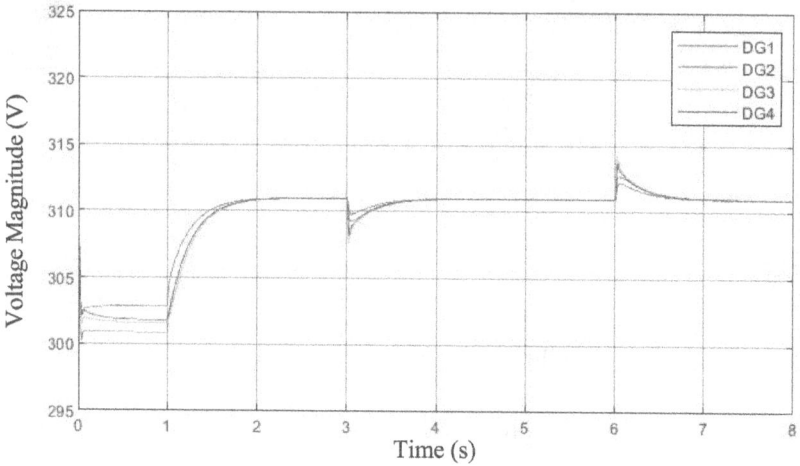

Figure 13.6   Voltage responses of DGs.

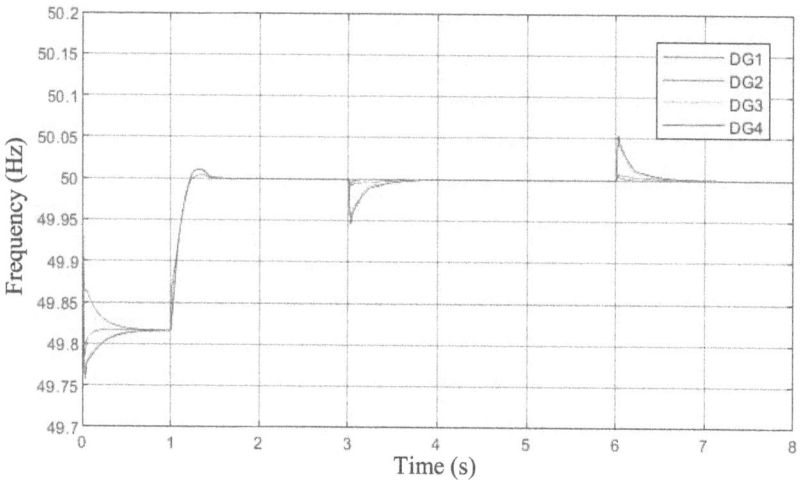

Figure 13.7   Frequency responses of DGs.

Figures 13.6–13.9 show the case study results. To be specific, Figure 13.6 and Figure 13.7 show the output voltage and frequency (VaF) of all DGs. The active power and its sharing performance are given in Figures 13.8–13.9. During $t = 0 \sim 1$s, although the primary control guarantees the stability of VaF, it inevitably brings VaF deviations from the nominal values for all DGs. After the droop control is stabilized, the voltage deviation of each DG is different, but the frequency deviation is the same. Once the proposed secondary control is implemented, the VaF are restored to their nominal values within 0.8 s and 0.5 s, respectively. The active power ratio $m_{P_i} P_i$

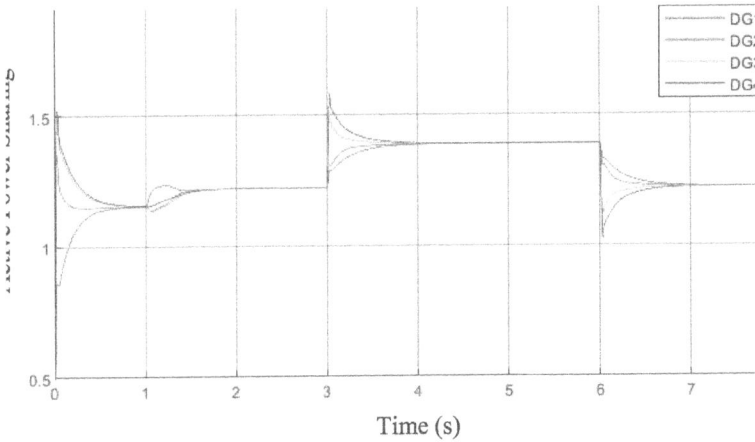

re 13.8 Active power sharing of DGs.

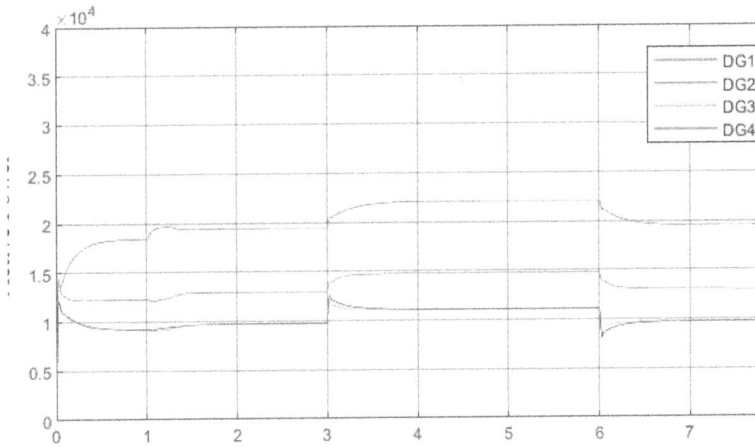

re 13.9 Active power responses of DGs.

; to be consistent within 0.5 s, i.e., a good active power sharing perfor
hieved. When the load changes at $t = 3$ s and $t = 6$ s, the VaF return t
ence values after small fluctuations. While each DG's active power out
es/decreases with the load changes, $m_{P_i}P_i$ tends to be consistent ensuring
e power sharing.

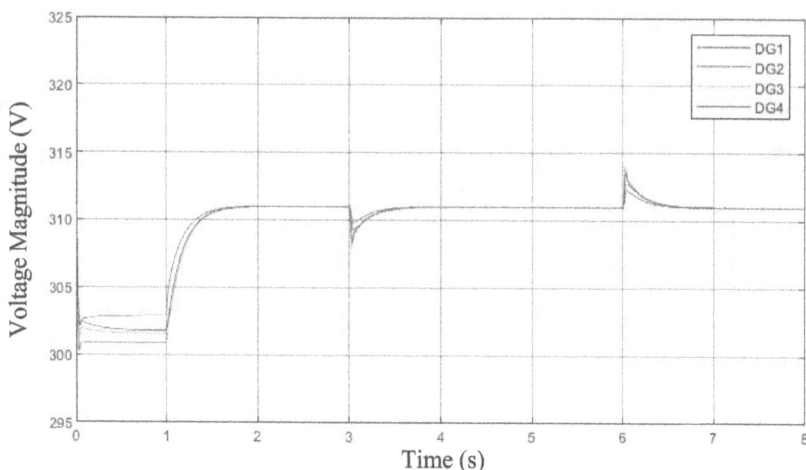

Figure 13.10 Voltage responses of DGs when C_{fi} and L_{fi} are uncertain (15% less than the rated values).

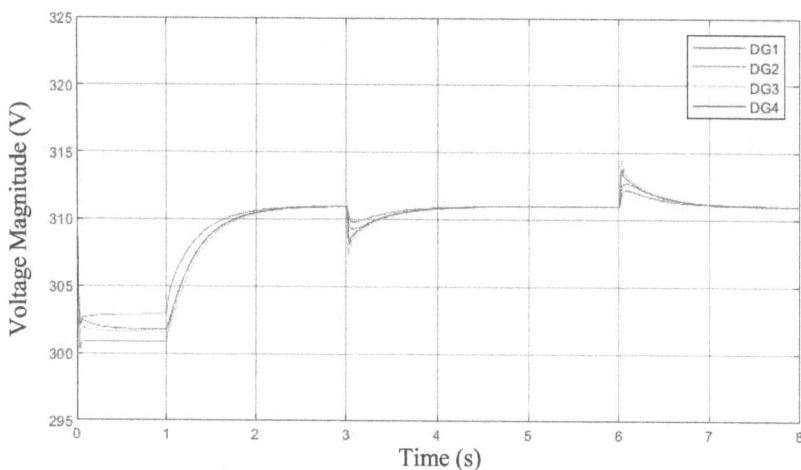

Figure 13.11 Voltage responses of DGs when C_{fi} and L_{fi} are uncertain (15% more than the rated values).

responses of the DGs are shown in Figures 13.10–13.11. It is observed that the proposed method achieves voltage restoration in the presence of parameter uncertainties. Compared with the voltage response using rated C_{fi} and L_{fi}, the output voltage only has a slight variation.

13.6 NOTES

This chapter has proposed a distributed robust secondary voltage and frequency control method for islanded microgrids. Each DG only requires the information of its neighbors for the controller design. In this method, the static gain control approach as well as a disturbance observer is used. It is shown that the proposed method improves the control performance and is robust against system uncertainties. Case studies have verified the efficacy of the proposed method under different conditions.

14 Conclusions and Future Challenges

14.1 CONCLUSIONS

This book aims to present innovative technologies for the design of variable gain controllers for nonlinear systems. It systematically describes the origin and recent results of variable gain control for nonlinear systems, focusing on controller design and stability analysis. In addition, the methods of variable gain control for energy conversion are also presented. Compared to the existing literature, the following research topics and novel solutions are presented in a series.

- **Variable gain control for different kinds of nonlinear systems**
  The theoretical part of this book deals with the control design of single nonlinear systems and large-scale interconnected nonlinear systems. These considered systems all possess a triangular structure, including strict-feedback nonlinear systems with a lower-triangular structure and feedforward nonlinear systems with an upper-triangular structure. Specifically, this book solves the control design problem for strict-feedback nonlinear systems (Chapters 2-4, 10) and feedforward nonlinear systems (Chapters 5-9). In addition, some extended forms of the above triangular systems are also considered, such as time-varying systems in Chapters 2 and 5, time-delayed systems in Chapters 3 and 5, and complex, large-scale interconnected systems in Chapters 8-10.
- **Different techniques for building control gains**
  This book uses the variable gain control method to study the control design of nonlinear systems. By introducing an appropriate state transformation, the problem of controller design can be transformed into the problem of designing the control gains. Variable gain control can be divided into three categories, namely static gain control (Chapters 6, 8, 9), time-varying gain control (Chapters 9, 10), and dynamic gain control (Chapters 2-5, 7, 8). As a special case of variable gain control, static gain control can only ensure semi-global stability for complex nonlinear systems (Chapter 8). For time-varying gain control, it can effectively deal with unknown parameters of the considered system, thereby improving the control performance (Chapter 9). Dynamic gain control is able to handle complex nonlinear dynamics in the system under consideration, providing global stability results (Chapters 2-5, 7, 8). Moreover, compared to the conventional forwarding and backstepping design methods, dynamic gain control can avoid the tedious controller iterative process and is therefore easy to implement in practice.

- **The applications of variable gain control in the field of energy conversion**

  This book also contains some examples of the application of variable gain control methods to energy systems: (1) Static and dynamic gain control methods are both applied in the AC/DC converter system. The control structure of these adopted methods includes a state transformation and a single-loop controller, which is simpler than the traditional dual-loop control structure (Chapter 11). (2) A fixed-time dynamic gain control method is proposed for the speed regulation of permanent magnet synchronous motors. Compared to the traditional PI method, the proposed fixed-time method provides faster and more accurate speed-tracking performance (Chapter 12). (3) A distributed robust secondary voltage and frequency control method for islanded microgrids is proposed. It is shown that the proposed method improves control performance and is robust to system uncertainties (Chapter 13).

14.2   FUTURE CHALLENGES

Although we have discussed the theory and application of variable gain control in detail, there are still some interesting problems that have not been fully explored. Some future research topics are suggested as follows:

- The design method of variable gain control is based on robust aspects, and the gains designed sometimes can be too large or too small, resulting in high peak performance or slow convergence rate. Therefore, it is desirable to design the control gains in an adaptive way which can achieve a balance between the high peak performance and slow convergence rate.
- Extension of the developed variable gain design methods to large-scale feedforward nonlinear time-delay systems.
- Output feedback control of large-scale feedforward nonlinear systems.
- Variable gain control for large-scale discrete-time nonlinear systems.
- Design of direct power control with variable gains of three-phase AC/DC converters under unbalanced grids or weak grids.
- Distributed robust secondary control based on variable gains for islanded microgrids with cyber attacks.
- Application of variable gain control to other practical systems, such as active power filters, DC microgrids, inductive motors, robotic systems, unmanned aerial vehicles (UAVs), etc.

References

1. Tarek Ahmed-Ali, Frédéric Mazenc, and Françoise Lamnabhi-Lagarrigue. Disturbance attenuation for discrete-time feedforward nonlinear systems. In *Stability and Stabilization of Nonlinear Systems*, pages 1–17. Springer, 1999.

2. In-Cheol Baik, Kyeong-Hwa Kim, and Myung-Joong Youn. Robust nonlinear speed control of pm synchronous motor using boundary layer integral sliding mode control technique. *IEEE Transactions on Control Systems Technology*, 8(1):47–54, 2000.

3. Sanjay P. Bhat and Dennis S. Bernstein. Finite-time stability of continuous autonomous systems. *SIAM Journal on Control and Optimization*, 38(3):751–766, 2000.

4. Ali Bidram, Ali Davoudi, Frank L. Lewis, and Josep M. Guerrero. Distributed cooperative secondary control of microgrids using feedback linearization. *IEEE Transactions on Power Systems*, 28(3):3462–3470, 2013.

5. Ali Bidram, Ali Davoudi, Frank L. Lewis, and Zhihua Qu. Secondary control of microgrids based on distributed cooperative control of multi-agent systems. *IET Generation, Transmission & Distribution*, 7(8):822–831, 2013.

6. Frede Blaabjerg. *Control of Power Electronic Converters and Systems*. Elsevier, 2013.

7. Juan Manuel Carrasco, Leopoldo Garcia Franquelo, Jan T. Bialasiewicz, Eduardo Galván, Ramón Carlos Portillo Guisado, M. A. Martin Prats, José Ignacio León, and Narciso Moreno-Alfonso. Power-electronic systems for the grid integration of renewable energy sources: A survey. *IEEE Transactions on Industrial Electronics*, 53(4):1002–1016, 2006.

8. Le Chang, Qing-Long Han, Xiaohua Ge, Chenghui Zhang, and Xianfu Zhang. On designing distributed prescribed finite-time observers for strict-feedback nonlinear systems. *IEEE Transactions on Cybernetics*, 51(9):4695–4706, 2021.

9. Le Chang, Chenghui Zhang, Xianfu Zhang, and Xiandong Chen. Decentralised regulation of nonlinear multi-agent systems with directed network topologies. *International Journal of Control*, 90(11):2338–2348, 2016.

10. C. L. Philip Chen, Guo-Xing Wen, Yan-Jun Liu, and Zhi Liu. Observer-based adaptive backstepping consensus tracking control for high-order nonlinear semi-strict-feedback multiagent systems. *IEEE Transactions on Cybernetics*, 46(7):1591–1601, 2015.

11. Weisheng Chen and Junmin Li. Decentralized output-feedback neural control for systems with unknown interconnections. *IEEE Transactions on Systems, Man, and Cybernetics, Part B (Cybernetics)*, 38(1):258–266, 2008.

12. Yangquan Chen, Zhiming Gong, and Changyun Wen. Analysis of a high-order iterative learning control algorithm for uncertain nonlinear systems with state delays. *Automatica*, 34(3):345–353, 1998.

13. Yangquan Chen, Changyun Wen, Jian-Xin Xu, and Mingxuan Sun. An initial state learning method for iterative learning control of uncertain time-varying systems. In *Proceedings of 35th IEEE Conference on Decision and Control*, volume 4, pages 3996–4001. IEEE, 1996.

14. Zhiyong Chen and Jie Huang. Global output feedback stabilization for uncertain nonlinear systems with output dependent incremental rate. In *Proceedings of the 2004 American Control Conference*, volume 4, pages 3047–3052. IEEE, 2004.

15. Kuang-Yao Cheng and Ying-Yu Tzou. Fuzzy optimization techniques applied to the design of a digital PMSM servo drive. *IEEE Transactions on Power Electronics*, 19(4):1085–1099, 2004.

16. Felix L. Chernous ko, Igor M. Ananievski, and Sergey A. Reshmin. *Control of Non-linear Dynamical Systems: Methods and Applications*. Springer Science & Business Media, 2008.

17. Armando W. Colombo, Stamatis Karnouskos, Okyay Kaynak, Yang Shi, and Shen Yin. Industrial cyberphysical systems: A backbone of the fourth industrial revolution. *IEEE Industrial Electronics Magazine*, 11(1):6–16, 2017.

18. Michael Defoort, Andrey Polyakov, Guillaume Demesure, Mohamed Djemai, and Kalyana Veluvolu. Leader-follower fixed-time consensus for multi-agent systems with unknown non-linear inherent dynamics. *IET Control Theory & Applications*, 9(14):2165–2170, 2015.

19. Nima Mahdian Dehkordi, Nasser Sadati, and Mohsen Hamzeh. Distributed robust finite-time secondary voltage and frequency control of islanded microgrids. *IEEE Transactions on Power Systems*, 32(5):3648–3659, 2017.

20. Derui Ding, Qing-Long Han, Zidong Wang, and Xiaohua Ge. A survey on model-based distributed control and filtering for industrial cyber-physical systems. *IEEE Transactions on Industrial Informatics*, 15(5):2483–2499, 2019.

21. Haibo Du, Chunjiang Qian, Yigang He, and Yingying Cheng. Global sampled-data output feedback stabilisation of a class of upper-triangular systems with input delay. *IET Control Theory & Applications*, 7(10):1437–1446, 2013.

22. Haibo Du, Chunjiang Qian, and Shihua Li. Global stabilization of a class of uncertain upper-triangular systems under sampled-data control. *International Journal of Robust and Nonlinear Control*, 23(6):620–637, 2012.

23. Haibo Du, Guanghui Wen, Yingying Cheng, and Jinhu Lü. Design and implementation of bounded finite-time control algorithm for speed regulation of permanent magnet synchronous motor. *IEEE Transactions on Industrial Electronics*, 68(3):2417–2426, 2021.

24. Gerardo Escobar, D. Chevreau, Romeo Ortega, and Eduardo Mendes. An adaptive passivity-based controller for a unity power factor rectifier. *IEEE Transactions on Control Systems Technology*, 9(4):637–644, 2001.

25. Baris Fidan, Youping Zhang, and Petros A. Ioannou. Adaptive control of a class of slowly time varying systems with modeling uncertainties. *IEEE Transactions on Automatic Control*, 50(6):915–920, 2005.

26. Cheng Fu, Chenghui Zhang, Guanguan Zhang, Jinqiu Song, Chen Zhang, and Bin Duan. Disturbance observer-based finite-time control for three-phase AC–DC converter. *IEEE Transactions on Industrial Electronics*, 69(9):5637–5647, 2022.

27. Xiaobin Gao, Daniel Liberzon, and Tamer Başar. On stability of nonlinear slowly time-varying and switched systems. In *2018 IEEE Conference on Decision and Control (CDC)*, pages 6458–6463. IEEE, 2018.

28. Xiaobin Gao, Daniel Liberzon, Ji Liu, and Tamer Başar. Unified stability criteria for slowly time-varying and switched linear systems. *Automatica*, 96:110–120, 2018.

29. Pudong Ge, Xiaobo Dou, Xiangjun Quan, Qinran Hu, Wanxing Sheng, Zaijun Wu, and Wei Gu. Extended-state-observer-based distributed robust secondary voltage and frequency control for an autonomous microgrid. *IEEE Transactions on Sustainable Energy*, 11(1):195–205, 2018.

30. Pudong Ge, Yue Zhu, Tim C. Green, and Fei Teng. Resilient secondary voltage control of islanded microgrids: An ESKBF-based distributed fast terminal sliding mode control approach. *IEEE Transactions on Power Systems*, 36(2):1059–1070, 2021.

31. Xiaohua Ge and Qing-Long Han. Consensus of multiagent systems subject to partially accessible and overlapping Markovian network topologies. *IEEE Transactions on Cybernetics*, 47(8):1807–1819, 2016.

32. Xiaohua Ge, Qing-Long Han, Xian-Ming Zhang, Lei Ding, and Fuwen Yang. Distributed event-triggered estimation over sensor networks: A survey. *IEEE Transactions on Cybernetics*, 50(3):1306–1320, 2019.

33. Yonghao Gui, Mingshen Li, Jinghang Lu, Saeed Golestan, Josep M. Guerrero, and Juan C. Vasquez. A voltage modulated DPC approach for three-phase PWM rectifier. *IEEE Transactions on Industrial Electronics*, 65(10):7612–7619, 2018.

34. Yonghao Gui, Xiongfei Wang, and Frede Blaabjerg. Vector current control derived from direct power control for grid-connected inverters. *IEEE Transactions on Power Electronics*, 34(9):9224–9235, 2018.

35. Vehbi C. Gungor, Bin Lu, and Gerhard P. Hancke. Opportunities and challenges of wireless sensor networks in smart grid. *IEEE Transactions on Industrial Electronics*, 57(10):3557–3564, 2010.

36. Fanghong Guo, Changyun Wen, Jianfeng Mao, and Yong-Duan Song. Distributed secondary voltage and frequency restoration control of droop-controlled inverter-based microgrids. *IEEE Transactions on Industrial Electronics*, 62(7):4355–4364, 2014.

37. Yi Guo, Weibiao Zhou, and Peter L. Lee. H_∞ control for a class of structured time-delay systems. *International Journal of Control*, 45(1):35–47, 2002.

38. Jack K. Hale and Sjoerd M. Verduyn Lunel. *Introduction to functional differential equations*, volume 99. Springer Science & Business Media, 2013.

39. Lennart Harnefors, Alejandro G Yepes, Ana Vidal, and Jesús Doval-Gandoy. Passivity-based controller design of grid-connected VSCs for prevention of electrical resonance instability. *IEEE Transactions on Industrial Electronics*, 62(2):702–710, 2014.

40. Huifen Hong, Wenwu Yu, Guanghui Wen, and Xinghuo Yu. Distributed robust fixed-time consensus for nonlinear and disturbed multiagent systems. *IEEE Transactions on Systems, Man, and Cybernetics: Systems*, 47(7):1464–1473, 2016.

41. Yiguang Hong, Jiangping Hu, and Linxin Gao. Tracking control for multi-agent consensus with an active leader and variable topology. *Automatica*, 42(7):1177–1182, 2006.

42. Jiabing Hu, Lei Shang, Yikang He, and Z. Q. Zhu. Direct active and reactive power regulation of grid-connected DC/AC converters using sliding mode control approach. *IEEE Transactions on Power Electronics*, 26(1):210–222, 2010.

43. Chang-Chun Hua, Xiu You, and Xin-Ping Guan. Leader-following consensus for a class of high-order nonlinear multi-agent systems. *Automatica*, 73:138–144, 2016.

44. Changchun Hua, Xinping Guan, and Peng Shi. Robust backstepping control for a class of time delayed systems. *IEEE Transactions on Automatic Control*, 50(6):894–899, 2005.

45. Changchun Hua, Yafeng Li, and Xinping Guan. Finite/fixed-time stabilization for nonlinear interconnected systems with dead-zone input. *IEEE Transactions on Automatic Control*, 62(5):2554–2560, 2017.

46. Jiangshuai Huang, Wei Wang, Changyun Wen, and Jing Zhou. Adaptive control of a class of strict-feedback time-varying nonlinear systems with unknown control coefficients. *Automatica*, 93:98–105, 2018.

47. Xianqing Huang, Wei Lin, and Bo Yang. Global finite-time stabilization of a class of uncertain nonlinear systems. *Automatica*, 41(5):881–888, 2005.

48. Zhong-Ping Jiang, I. Mareels, D. J. Hill, and Jie Huang. A unifying framework for global regulation via nonlinear output feedback: From ISS to iISS. *IEEE Transactions on Automatic Control*, 49(4):549–562, 2004.

49. Iasson Karafyllis and Zhong-Ping Jiang. *Stability and Stabilization of Nonlinear Systems*. Springer Science & Business Media, 2011.

50. Hassan K. Khalil. *Nonlinear Systems; 3rd ed.* Prentice-Hall, Upper Saddle River, NJ, 2002.

51. Hassan K. Khalil. *High-Gain Observers in Nonlinear Feedback Control*. Society for Industrial and Applied Mathematics, 2017.

52. Yousef Khayat, Qobad Shafiee, Rasool Heydari, Mobin Naderi, Tomislav Dragičević, John W. Simpson-Porco, Florian Dörfler, Mohammad Fathi, Frede Blaabjerg, Josep M. Guerrero, et al. On the secondary control architectures of AC microgrids: An overview. *IEEE Transactions on Power Electronics*, 35(6):6482–6500, 2019.

53. Suiyang Khoo, Juliang Yin, Zhihong Man, and Xinghuo Yu. Finite-time stabilization of stochastic nonlinear systems in strict-feedback form. *Automatica*, 49(5):1403–1410, 2013.

54. Min-Sung Koo and Ho-Lim Choi. Non-predictor controller for feedforward and non-feedforward nonlinear systems with an unknown time-varying delay in the input. *Automatica*, 65:27–35, 2016.

55. Min-Sung Koo, Ho-Lim Choi, and Jong-Tae Lim. Global regulation of a class of feedforward and non-feedforward nonlinear systems with a delay in the input. *Automatica*, 48(10):2607–2613, 2012.

56. Prashanth Krishnamurthy and Farshad Khorrami. A high-gain scaling technique for adaptive output feedback control of feedforward systems. *IEEE Transactions on Automatic Control*, 49(12):2286–2292, 2004.

57. Prashanth Krishnamurthy and Farshad Khorrami. High-gain output-feedback control for nonlinear systems based on multiple time scaling. *Systems & Control Letters*, 56(1):7–15, 2007.

58. Prashanth Krishnamurthy and Farshad Khorrami. Feedforward systems with ISS appended dynamics: Adaptive output-feedback stabilization and disturbance attenuation. *IEEE Transactions on Automatic Control*, 53(1):405–412, 2008.

59. Miroslav Krstic, Petar V. Kokotovic, and Ioannis Kanellakopoulos. *Nonlinear and Adaptive Control Design*. John Wiley & Sons, Inc., 1995.

60. Hao Lei and Wei Lin. Adaptive regulation of uncertain nonlinear systems by output feedback: A universal control approach. *Systems & Control Letters*, 56(7-8):529–537, 2007.

61. Jing Lei and Hassan K. Khalil. High-gain-predictor-based output feedback control for time-delay nonlinear systems. *Automatica*, 71:324–333, 2016.

62. Hanfeng Li, Xianfu Zhang, and Qingrong Liu. Adaptive output feedback control for a class of large-scale output-constrained non-linear time-delay systems. *IET Control Theory & Applications*, 12(1):174–181, 2018.

63. Hanfeng Li, Xianfu Zhang, and Shuai Liu. An improved dynamic gain method to global regulation of feedforward nonlinear systems. *IEEE Transactions on Automatic Control*, 67(6):2981–2988, 2022.

64. Jian Li and Dezhen Zhang. Backstepping and sliding-mode techniques applied to distributed secondary control of islanded microgrids. *Asian Journal of Control*, 20(3):1288–1295, 2018.

65. Junpeng Li, Yana Yang, Changchun Hua, and Xinping Guan. Fixed-time backstepping control design for high-order strict-feedback non-linear systems via terminal sliding mode. *IET Control Theory & Applications*, 11(8):1184–1193, 2017.

66. Kuo Li, Chang-Chun Hua, Xiu You, and Xin-Ping Guan. Output feedback-based consensus control for nonlinear time delay multiagent systems. *Automatica*, 111:108669, 2020.

67. Shihua Li, Mingming Zhou, and Xinghuo Yu. Design and implementation of terminal sliding mode control method for PMSM speed regulation system. *IEEE Transactions on Industrial Informatics*, 9(4):1879–1891, 2012.

68. Weixun Li and Zengqiang Chen. Leader-following consensus of second-order multi-agent systems with time-delay and nonlinear dynamics. In *Proceedings of the 31st Chinese Control Conference*, pages 6124–6128. IEEE, 2012.

69. Wuquan Li, Xiaoxiao Yao, and Miroslav Krstic. Adaptive-gain observer-based stabilization of stochastic strict-feedback systems with sensor uncertainty. *Automatica*, 120:109112, 2020.

70. Xiao-Jian Li and Guang-Hong Yang. Adaptive decentralized control for a class of interconnected nonlinear systems via backstepping approach and graph theory. *Automatica*, 76:87–95, 2017.

71. Zhongkui Li, Xiangdong Liu, Mengyin Fu, and Lihua Xie. Global H_∞ consensus of multi-agent systems with Lipschitz non-linear dynamics. *IET Control Theory & Applications*, 6(13):2041–2048, 2012.

72. Zhongkui Li, Wei Ren, Xiangdong Liu, and Lihua Xie. Distributed consensus of linear multi-agent systems with adaptive dynamic protocols. *Automatica*, 49(7):1986–1995, 2013.

73. Zhongwen Li, Chuanzhi Zang, Peng Zeng, Haibin Yu, and Hepeng Li. MAS based distributed automatic generation control for cyber-physical microgrid system. *IEEE/CAA Journal of Automatica Sinica*, 3(1):78–89, 2016.

74. Zongli Lin. Robust semi-global stabilization of linear systems with imperfect actuators. *Systems & Control Letters*, 29(4):215–221, 1997.

75. Zongli Lin. Almost disturbance decoupling with global asymptotic stability for nonlinear systems with disturbance-affected unstable zero dynamics. *Systems & Control Letters*, 33(3):163–169, 1998.

76. Zongli Lin. *Low Gain Feedback*. Springer London, 1999.

77. Zongli Lin. Low gain and low-and-high gain feedback: A review and some recent results. In *2009 Chinese Control and Decision Conference*. IEEE, 2009.

78. Huixian Liu and Shihua Li. Speed control for PMSM servo system using predictive functional control and extended state observer. *IEEE Transactions on Industrial Electronics*, 59(2):1171–1183, 2011.

79. Jianxing Liu, Sergio Vazquez, Ligang Wu, Abraham Marquez, Huijun Gao, and Leopoldo G. Franquelo. Extended state observer-based sliding-mode control for three-phase power converters. *IEEE Transactions on Industrial Electronics*, 64(1):22–31, 2016.

80. Qingrong Liu and Zhishan Liang. Finite-time consensus of time-varying nonlinear multi-agent systems. *International Journal of Systems Science*, 47(11):2642–2651, 2016.

81. Shuai Liu, Lihua Xie, and Frank L. Lewis. Synchronization of multi-agent systems with delayed control input information from neighbors. *Automatica*, 47(10):2152–2164, 2011.

82. Shutang Liu, Weiyong Yu, and Fangfang Zhang. Output feedback regulation for large-scale uncertain nonlinear systems with time delays. *Kybernetika*, 51(5):874–889, 2015.

83. Wei Liu and Jie Huang. Adaptive leader-following consensus for a class of higher-order nonlinear multi-agent systems with directed switching networks. *Automatica*, 79:84–92, 2017.

84. Xudong Liu, Haisheng Yu, Jinpeng Yu, and Yang Zhao. A novel speed control method based on port-controlled Hamiltonian and disturbance observer for PMSM drives. *IEEE Access*, 7:111115–111123, 2019.

85. Xudong Liu, Chenghui Zhang, Ke Li, and Qi Zhang. Robust current control-based generalized predictive control with sliding mode disturbance compensation for PMSM drives. *ISA Transactions*, 71:542–552, 2017.

86. Yungang Liu. Global finite-time stabilization via time-varying feedback for uncertain nonlinear systems. *SIAM Journal on Control and Optimization*, 52(3):1886–1913, 2014.

87. Francisco Lopez-Ramirez, Andrey Polyakov, Denis Efimov, and Wilfrid Perruquetti. Finite-time and fixed-time observer design: Implicit Lyapunov function approach. *Automatica*, 87:52–60, 2018.

88. Languang Lu, Xuebing Han, Jianqiu Li, Jianfeng Hua, and Minggao Ouyang. A review on the key issues for lithium-ion battery management in electric vehicles. *Journal of Power Sources*, 226:272–288, 2013.

89. Wensheng Luo, Yunfei Yin, Xiangyu Shao, Jianxing Liu, and Ligang Wu. *Advanced Control Methodologies for Power Converter Systems*. Springer, 2022.

90. Cui-Qin Ma and Ji-Feng Zhang. Necessary and sufficient conditions for consensusability of linear multi-agent systems. *IEEE Transactions on Automatic Control*, 55(5):1263–1268, 2010.

91. F. Mazenc, S. Mondie, and R. Francisco. Global asymptotic stabilization of feedforward systems with delay in the input. In *42nd IEEE International Conference on Decision and Control (IEEE Cat. No. 03CH37475)*. IEEE, December 2003.

92. F. Mazenc and Henk Nijmeijer. Forwarding in discrete-time nonlinear systems. *International Journal of Control*, 71(5):823–835, 1998.

93. F. Mazenc, L. Praly, and W. P. Dayawansa. Global stabilization by output feedback: examples and counterexamples. *Systems & Control Letters*, 23(2):119–125, 1994.

94. Frederic Mazenc, Sabine Mondie, and Rogelio Francisco. Global asymptotic stabilization of feedforward systems with delay in the input. *IEEE Transactions on Automatic Control*, 49(5):844–850, 2004.

95. Frederic Mazenc, Sabine Mondie, and Silviu-Iulian Niculescu. Global asymptotic stabilization for chains of integrators with a delay in the input. *IEEE Transactions on Automatic Control*, 48(1):57–63, 2003.

96. Frédéric Mazenc, Silviu-Iulian Niculescu, and Mounir Bekaik. Stabilization of time-varying nonlinear systems with distributed input delay by feedback of plant's state. *IEEE Transactions on Automatic Control*, 58(1):264–269, 2012.

97. Aniruddh Mohan, Shashank Sripad, Parth Vaishnav, and Venkatasubramanian Viswanathan. Trade-offs between automation and light vehicle electrification. *Nature Energy*, 5(7):543–549, 2020.

98. Boda Ning, Qing-long Han, and Lei Ding. Distributed secondary control of ac microgrids with external disturbances and directed communication topologies: A full-order sliding-mode approach. *IEEE/CAA Journal of Automatica Sinica*, 8(3):554–564, 2021.

99. Boda Ning, Qing-Long Han, and Zongyu Zuo. Practical fixed-time consensus for integrator-type multi-agent systems: A time base generator approach. *Automatica*, 105:406–414, 2019.

100. Boda Ning and Han Qing-Long. Prescribed finite-time consensus tracking for multiagent systems with nonholonomic chained-form dynamics. *IEEE Transactions on Automatic Control*, 64(4):1686–1693, 2019.

101. Andrey Polyakov. Nonlinear feedback design for fixed-time stabilization of linear control systems. *IEEE Transactions on Automatic Control*, 57(8):2106–2110, 2012.

102. L. Praly. Asymptotic stabilization via output feedback for lower triangular systems with output dependent incremental rate. *IEEE Transactions on Automatic Control*, 48(6):1103–1108, 2003.

103. Laurent Praly and Zhong-Ping Jiang. Linear output feedback with dynamic high gain for nonlinear systems. *Systems & Control Letters*, 53(2):107–116, 2004.

104. Liang Qi and Hongbo Shi. Adaptive position tracking control of permanent magnet synchronous motor based on RBF fast terminal sliding mode control. *Neurocomputing*, 115:23–30, 2013.

105. Chunjiang Qian and Wei Lin. Using small feedback to stabilize a wider class of feedforward systems. *IFAC Proceedings Volumes*, 32(2):2434–2439, 1999.

106. Chunjiang Qian and Wei Lin. Output feedback control of a class of nonlinear systems: A nonseparation principle paradigm. *IEEE Transactions on Automatic Control*, 47(10):1710–1715, October 2002.

107. Jiahu Qin, Huijun Gao, and Wei Xing Zheng. Second-order consensus for multi-agent systems with switching topology and communication delay. *Systems & Control Letters*, 60(6):390–397, 2011.

108. F. H. Ramirez-Leyva, E Peralta-Sánchez, J. J. Vásquez-Sanjuan, and F. Trujillo-Romero. Passivity-based speed control for permanent magnet motors. *Procedia Technology*, 7:215–222, 2013.

109. Wei Ren and Randal W. Beard. *Distributed Consensus in Multi-Vehicle Cooperative Control*. Springer, 2008.

110. Vasso Reppa, Marios M Polycarpou, and Christos G Panayiotou. Distributed sensor fault diagnosis for a network of interconnected cyberphysical systems. *IEEE Transactions on Control of Network Systems*, 2(1):11–23, 2015.

111. Jean-Pierre Richard. Time-delay systems: An overview of some recent advances and open problems. *Automatica*, 39(10):1667–1694, 2003.

112. Wilson J. Rugh and Jeff S. Shamma. Research on gain scheduling. *Automatica*, 36(10):1401–1425, 2000.

113. Jin Heon Seo, Hyungbo Shim, and Juhoon Back. Consensus of high-order linear systems using dynamic output feedback compensator: Low gain approach. *Automatica*, 45(11):2659–2664, 2009.

114. Yunlong Shang. Optimization design and implementation of state estimation and balancing management system for Lithium-Ion batteries in electric vehicles. *Doctoral dissertation, Jinan: Shandong University*, 2017.

115. Yang Shi and Bo Yu. Output feedback stabilization of networked control systems with random delays modeled by Markov chains. *IEEE Transactions on Automatic Control*, 54(7):1668–1674, 2009.

116. Bhim Singh, Brij N. Singh, Ambrish Chandra, Kamal Al-Haddad, Ashish Pandey, and Dwarka P. Kothari. A review of three-phase improved power quality AC-DC converters. *IEEE Transactions on Industrial Electronics*, 51(3):641–660, 2004.

117. Yongduan Song, Yujuan Wang, John Holloway, and Miroslav Krstic. Time-varying feedback for regulation of normal-form nonlinear systems in prescribed finite time. *Automatica*, 83:243–251, 2017.

118. Shize Su and Zongli Lin. Distributed consensus control of multi-agent systems with higher order agent dynamics and dynamically changing directed interaction topologies. *IEEE Transactions on Automatic Control*, 61(2):515–519, 2015.

119. Shize Su, Yusheng Wei, and Zongli Lin. Stabilization of discrete-time linear systems with an unknown time-varying delay by switched low-gain feedback. *IEEE Transactions on Automatic Control*, 64(5):2069–2076, 2019.

120. Xiaofei Sun, Haisheng Yu, Jinpeng Yu, and Xudong Liu. Design and implementation of a novel adaptive backstepping control scheme for a PMSM with unknown load torque. *IET Electric Power Applications*, 13(4):445–455, 2019.

121. Yumei Sun, Fang Wang, Zhi Liu, Yun Zhang, and C. L. Philip Chen. Fixed-time fuzzy control for a class of nonlinear systems. *IEEE Transactions on Cybernetics*, 52(5):3880–3887, 2022.

122. A. R. Teel. Using saturation to stabilize a class of single-input partially linear composite systems. *IFAC Proceedings Volumes*, 25(13):379–384, June 1992.

123. Sergio Vazquez, Jose Rodriguez, Marco Rivera, Leopoldo G Franquelo, and Margarita Norambuena. Model predictive control for power converters and drives: Advances and trends. *IEEE Transactions on Industrial Electronics*, 64(2):935–947, 2016.

124. Rong-Jong Wai and Yan Yang. Design of backstepping direct power control for three-phase PWM rectifier. *IEEE Transactions on Industry Applications*, 55(3):3160–3173, 2019.

125. Jinhuan Wang, Daizhan Cheng, and Xiaoming Hu. Consensus of multi-agent linear dynamic systems. *Asian Journal of Control*, 10(2):144–155, 2008.

126. Lei Wang, Daniele Astolfi, Lorenzo Marconi, and Hongye Su. High-gain observers with limited gain power for systems with observability canonical form. *Automatica*, 75:16–23, 2017.

127. Xing-Hu Wang and Hai-Bo Ji. Leader-follower consensus for a class of nonlinear multi-agent systems. *International Journal of Control, Automation and Systems*, 10(1):27–35, 2012.

128. Yujuan Wang, Yongduan Song, David J. Hill, and Miroslav Krstic. Prescribed-time consensus and containment control of networked multiagent systems. *IEEE Transactions on Cybernetics*, 49(4):1138–1147, 2019.

129. Yusheng Wei and Zongli Lin. On the delay bounds of discrete-time linear systems under delay independent truncated predictor feedback. In *2016 American Control Conference (ACC)*, pages 89–94. IEEE, 2016.

130. Changyun Wen, Jing Zhou, Zhitao Liu, and Hongye Su. Robust adaptive control of uncertain nonlinear systems in the presence of input saturation and external disturbance. *IEEE Transactions on Automatic Control*, 56(7):1672–1678, 2011.

131. Guoxing Wen, C. L. Philip Chen, Yan-Jun Liu, and Zhi Liu. Neural network-based adaptive leader-following consensus control for a class of nonlinear multiagent state-delay systems. *IEEE Transactions on Cybernetics*, 47(8):2151–2160, 2016.

132. Lantao Xing, Changyun Wen, Zhitao Liu, Hongye Su, and Jianping Cai. Event-triggered adaptive control for a class of uncertain nonlinear systems. *IEEE Transactions on Automatic Control*, 62(4):2071–2076, 2016.

133. Shengyuan Xu and Paul Van Dooren. Robust H_∞ filtering for a class of non-linear systems with state delay and parameter uncertainty. *International Journal of Control*, 37(3):141–148, 2006.

134. Hao Yang, Bin Jiang, Marcel Staroswiecki, and Youmin Zhang. Fault recoverability and fault tolerant control for a class of interconnected nonlinear systems. *Automatica*, 54:49–55, 2015.

135. Xuefei Yang and Bin Zhou. Global stabilization of discrete-time feedforward time-delay systems by bounded controls. *International Journal of Robust and Nonlinear Control*, 28(15):4438–4454, 2018.

136. Huawen Ye. Saturated delayed controls for feedforward nonlinear systems. *IEEE Transactions on Automatic Control*, 59(6):1646–1653, 2013.

137. X. Ye and H. Unbehauen. Global adaptive stabilization for a class of feedforward nonlinear systems. *IEEE Transactions on Automatic Control*, 49(5):786–792, 2004.

138. Xudong Ye. Universal stabilization of feedforward nonlinear systems. *Automatica*, 39(1):141–147, 2003.

139. Yunfei Yin, Jianxing Liu, Wensheng Luo, Ligang Wu, Sergio Vazquez, José I. Leon, and Leopoldo G Franquelo. Adaptive control for three-phase power converters with disturbance rejection performance. *IEEE Transactions on Systems, Man, and Cybernetics: Systems*, 51(2):674–685, 2018.

140. Yunfei Yin, Sergio Vazquez, Abraham Marquez, Jianxing Liu, José I. Leon, Ligang Wu, and Leopoldo G. Franquelo. Observer-based sliding-mode control for grid-connected power converters under unbalanced grid conditions. *IEEE Transactions on Industrial Electronics*, 69(1):517–527, 2021.

141. Sung Jin Yoo. Distributed adaptive containment control of uncertain nonlinear multi-agent systems in strict-feedback form. *Automatica*, 49(7):2145–2153, 2013.

142. Jinpeng Yu, Peng Shi, Wenjie Dong, Bing Chen, and Chong Lin. Neural network-based adaptive dynamic surface control for permanent magnet synchronous motors. *IEEE Transactions on Neural Networks and Learning Systems*, 26(3):640–645, 2014.

143. Tianyi Zeng, Xuemei Ren, and Yao Zhang. Fixed-time sliding mode control and high-gain nonlinearity compensation for dual-motor driving system. *IEEE Transactions on Industrial Informatics*, 16(6):4090–4098, 2019.

144. Chenghui Zhang, Le Chang, and Xianfu Zhang. Leader-follower consensus of upper-triangular nonlinear multi-agent systems. *IEEE/CAA Journal of Automatica Sinica*, 1(2):210–217, 2014.

145. Chenghui Zhang, Ke Li, Naxin Cui, Guojing Xing, Jian Wu, and Bo Sun. Research progress on key control problems arising from the energy and driving system of the hybrid electric vehicle. *Journal of Shandong University (Engineering Science)*, 41(5):1–8, 2011.

146. Kemei Zhang and Xing-Hui Zhang. Finite-time stabilisation for high-order nonlinear systems with low-order and high-order nonlinearities. *International Journal of Control*, 88(8):1576–1585, 2015.

147. Xianfu Zhang, Luc Baron, Qingrong Liu, and El-Kebir Boukas. Design of stabilizing controllers with a dynamic gain for feedforward nonlinear time-delay systems. *IEEE Transactions on Automatic Control*, 56(3):692–697, 2011.

148. Xianfu Zhang, Gang Feng, and Yonghui Sun. Finite-time stabilization by state feedback control for a class of time-varying nonlinear systems. *Automatica*, 48(3):499–504, 2012.

149. Xianfu Zhang, Hongyan Gao, and Chenghui Zhang. Global asymptotic stabilization of feedforward nonlinear systems with a delay in the input. *International Journal of Systems Science*, 37(3):141–148, 2006.

150. Xianfu Zhang, Lu Liu, and Gang Feng. Leader–follower consensus of time-varying nonlinear multi-agent systems. *Automatica*, 52:8–14, 2015.

151. Xianfu Zhang, Lu Liu, Gang Feng, and Chenghui Zhang. Output feedback control of large-scale nonlinear time-delay systems in lower triangular form. *Automatica*, 49(11):3476–3483, 2013.

152. Xianfu Zhang, Chenghui Zhang, and Zhaolin Cheng. Asymptotic stabilization via output feedback for nonlinear systems with delayed output. *International Journal of Systems Science*, 37(9):599–607, 2006.

153. Xianfu Zhang, Chenghui Zhang, and Yuzhen Wang. Decentralized output feedback stabilization for a class of large-scale feedforward nonlinear time-delay systems. *International Journal of Robust and Nonlinear Control*, 24(17):2628–2639, 2014.

154. Xu Zhang, Wei Lin, and Yan Lin. Nonsmooth feedback control of time-delay nonlinear systems: A dynamic gain based approach. *IEEE Transactions on Automatic Control*, 62(1):438–444, 2016.

155. Xu Zhang and Yan Lin. Adaptive output feedback control for a class of large-scale nonlinear time-delay systems. *Automatica*, 52:87–94, 2015.

156. Xueji Zhang, Kristian Hengster-Movrić, Michael Sebek, Wim Desmet, and Cassio Faria. Distributed observer and controller design for spatially interconnected systems. *IEEE Transactions on Control Systems Technology*, 27(1):1–13, 2017.

157. Feng Zheng and Paul M. Frank. Robust control of uncertain distributed delay systems with application to the stabilization of combustion in rocket motor chambers. *Automatica*, 38(3):487–497, 2002.

158. Bin Zhou, Zongli Lin, and Guang-Ren Duan. A parametric Lyapunov equation approach to low gain feedback design for discrete-time systems. *Automatica*, 45(1):238–244, 2009.

159. Bin Zhou, Wim Michiels, and Jie Chen. Fixed-time stabilization of linear delay systems by smooth periodic delayed feedback. *IEEE Transactions on Automatic Control*, 67(2):557–573, 2022.

160. Bin Zhou, Qian Wang, Zongli Lin, and Guang-Ren Duan. Gain scheduled control of linear systems subject to actuator saturation with application to spacecraft rendezvous. *IEEE Transactions on Control Systems Technology*, 22(5):2031–2038, 2014.

161. Bin Zhou and Xuefei Yang. Global stabilization of feedforward nonlinear time-delay systems by bounded controls. *Automatica*, 88:21–30, 2018.

162. Quanxin Zhu and Hui Wang. Output feedback stabilization of stochastic feedforward systems with unknown control coefficients and unknown output function. *Automatica*, 87:166–175, 2018.

163. Wei Zhu and Daizhan Cheng. Leader-following consensus of second-order agents with multiple time-varying delays. *Automatica*, 46(12):1994–1999, 2010.

164. Wei Zhu and Zhong-Ping Jiang. Event-based leader-following consensus of multi-agent systems with input time delay. *IEEE Transactions on Automatic Control*, 60(5):1362–1367, 2014.

165. Kunyu Zuo and Lei Wu. A review of decentralized and distributed control approaches for islanded microgrids: Novel designs, current trends, and emerging challenges. *The Electricity Journal*, 35(5):107138, 2022.

166. Shan Zuo, Ali Davoudi, Yongduan Song, and Frank L. Lewis. Distributed finite-time voltage and frequency restoration in islanded ac microgrids. *IEEE Transactions on Industrial Electronics*, 63(10):5988–5997, 2017.

167. Zongyu Zuo, Qing-Long Han, and Boda Ning. *Fixed-Time Cooperative Control of Multi-Agent Systems*. Springer International Publishing, 2019.

Index

For Product Safety Concerns and Information please contact our EU
representative GPSR@taylorandfrancis.com
Taylor & Francis Verlag GmbH, Kaufingerstraße 24, 80331 München, Germany